HMCS SACKVILLE
1941 - 1985

The Canadian Naval Memorial Trust
Le Fonds de commemoration de la marine canadienne
HMCS Sackville
Halifax, Nova Scotia

HMCS SACKVILLE
1941-1985

MARC MILNER

The Canadian Naval Memorial Trust
Le Fonds de commemoration
de la marine canadienne
HMCS Sackville
Halifax, Nova Scotia

To those who serve in war and peace

Copyright© 1998 by Marc Milner. All rights reserved. No part of this publication may be reproduced, stored in a retrieval system, or transmitted in any form or by any means, without the prior written permission of the publisher.

All proceeds from the sale of this book are dedicated to the Canadian Naval Memorial Trust.

Design: Linda Moroz-Irvine
Painting on cover reproduced with the kind permission of Robert Meecham

The Canadian Naval Memorial Trust
HMCS Sackville
FMO Halifax
Halifax, Nova Scotia
Canada B3K 2X0

Printed in Canada

02 01 00 99 98   5 4 3 2 1

Canadian Cataloguing in Publication Data

Milner, Marc
    HMCS Sackville 1941-1985

ISBN 0-9683661-0-4

1. Sackville (Ship) – History. I. Canadian Naval Memorial Trust. II. Title.

V825.5.C3M54 1998   359.3'254'0971   C98-930998-3

TABLE OF CONTENTS

Foreword . . . .*vi*

Acknowledgements . . . .*vii*

Chapter 1:
THE LONG ROAD TO WAR. 9

Chapter 2:
BATTLING THE WOLF PACKS 23

Chapter 3:
FROM THE GNAT TO GALVESTON 39

Chapter 4:
THE END OF ONE WAR
AND THE START OF ANOTHER 51

Chapter 5:
OCEANOGRAPHIC SERVICE. 61

Chapter 6:
RESTORATION 75

Appendix:
CORVETTES IN CANADIAN SERVICE 93

FOREWORD

Here is a concise account of a Canadian ship of war and peace written by a historian of repute, particularly for his narratives of our naval battles in the Atlantic of half a century ago. Marc Milner has blended in text and illustration the long career of HMCS *Sackville* in such a way that she reveals her versatility and, in a way, her character.

Built as a corvette she struggled with many others like her, sometimes half naked for want of adequate armament, to keep the sealanes open for convoys to pass unmolested. The ship had a personality; she responded to an order as an affectionate dog answers her master's call. She guarded her flock and drove off the wolves in time of war, mounted high seas with canine confidence and would lie patiently in some remote bay while the hydrographer did his work.

Curiously she did not fade away as others did when the war was over, she simply changed her coat and became an auxiliary vessel removing harbour defences then, changing once more to become a government research vessel destined to sail for years along the eastern seaboard from the Bay of Fundy to the Arctic on a variety of marine scientific studies.

Now HMCS *Sackville*, dressed in her original colours and fitted with the weapons of her past – a transformation clearly illustrated and described by Milner – rests at her own pier in Halifax Harbour in summertime for all to see and pace her decks, a national monument, a Canadian naval memorial.

Lt Cdr Alan Easton, RCN (R)
Captain, HMCS *Sackville*, 1942-43

ACKNOWLEDGEMENTS

Much of the preliminary work for this history was assembled by members of the Memorial Trust, to whom I extend my thanks and appreciation. Thanks also to the Trust members who made the book itself a reality, among them Fred Crickard, Charles Westropp, Vern Howland, Ted Smith, Maurice McGaffney, and Bill Gard.

Roger Sarty and Mike Whitby, at the Directorate of History and Heritage, NDHQ provided their usual unfailing support. Barb Tenholme at DREA, helped track down *Sackville*'s history as a research vessel, Harold Merklinger, DREA, answered questions, read the draft and provided photos. Morgan Smith from the MARCOM Museum, David Flemming of the Maritime Museum of the Atlantic, the staff at the National Photograph Collection, Ottawa, and the ever-reliable Ken Macpherson helped with the photos. A special thanks goes to Tony German for sharing his research on the mutiny in early 1942, and to Angela Dobler at Vanwell for her conscientious editing and enthusiasm for the project. If I have missed someone from this all-too-brief list, please accept my apologies. Any errors or omissions that remain in the text are mine.

And finally two special acknowledgements. The first is to Alan Easton for writing the foreword. His classic memoir *50 North* has immortalized *Sackville*'s 1942 exploits and, as the first of the ship's two great wartime captains, his association with this project is particularly gratifying. Finally, I would like to thank the Trust itself for allowing me the opportunity to write this book. A a child I was utterly fascinated by the knowledge that a warship named after my hometown had fought with such distinction in "my father's war." That fascination blossomed into a career as a naval historian and so, in no small way, *Sackville* shaped my destiny, too. It is, therefore, especially pleasing to use that professional skill to do something in return for the ship.

Marc Milner

Saint John Shipbuilding and Drydock Company, from the air in December 1942. The building slips on the left, by now occupied by merchant ships under construction, were used to build *Sackville* and her sisters. (NAC PA 197028)

CHAPTER 1

THE LONG ROAD TO WAR

The small group of dignitaries, dockyard workers and locals were lashed by a driving rain as they waited for Mrs. J.E.W. Oland to christen the new ship and send her down the ways into Courtenay Bay. It had taken nearly eleven months to get Patrol Vessel 2 (PV2) ready to launch, and she was still a long way from being ready for war.

Saint John Shipbuilding and Drydock Company had taken on the building of three corvettes in early 1940 when things were slow. But when Europe fell to the Germans that spring the yard was soon buried in the emergency repair work that kept it busy for the rest of the war. PVs 1 and 3 on nearby slips had also been delayed. However on that cold and rain-soaked 15 May 1941, PV 2, one of the most remarkable ships in Canadian history, was at last ready to hit the water.

As Mrs. Oland, the ship's sponsor, drew back the bottle of champagne everyone strained to hear her voice through the pelting rain. "God Bless His Majesty's Canadian Ship *Sackville* and all who sail in her!" she said finally. The bottle smashed against the ship's bow, the final blocks were knocked away and *Sackville* slipped stern first into the bay. Mayor Norman Hesler and the entire town council from Sackville, New Brunswick, were there to watch her go. So too were many others from vantage points along the shore, and – according to press reports – the new corvette was "greeted raucously by the sirens and whistles of other vessels in the harbour." As dockyard workers scattered to recover pieces of the ribbon and bunting that adorned the ship, tugs were already nudging her towards the berth where the long fitting-out process would take place.

Sackville was the second Flower Class corvette ordered for the Royal Canadian Navy (RCN) during the Second World War, but when she finally hit the water in May 1941 the RCN already had thirty-three similar ships in commission and nearly fifty more nearing completion. By the time the war ended in 1945, 123 corvettes of various types had joined the Canadian fleet and like *Sackville* most were built in Canadian yards. They were the largest class of warships ever to serve Canada, and the Canadian corvette fleet – part of some 269 corvettes built during the war by the Allies – was a crucial component of Allied victory in the battle of the Atlantic.

Of those 269 small ships that made Allied victory in the Atlantic possible, *Sackville* is the last known survivor.

The corvette was a British design, prepared by Smith's Dock Company Limited, and based on that yard's recent whale-catcher *Southern Pride*. Since these were to be mass produced auxiliary vessels manned in war by reservists, the ship was simple to build and operate. *Southern Pride*'s basic hull shape was extended to 205 feet overall and changed slightly to allow a gun to be mounted forward, which brought displacement up to 940 tons. A few modifications were made to the superstructure, internal layout, communications equipment and accommodation. The Royal Navy dubbed the new ships "Flower Class Corvettes" and gave them all appropriate names like *Hybiscus* and *Poppy*. The first British corvettes were ordered in July 1939, the same month that plans for the vessels arrived in Canada.

Initially the RCN had no use for these British "patrol vessels". They were designed to mercantile standards (in terms of

Early days: *Sackville* begins to take shape on the right, while the frames of *Amherst*'s foc'sle are in evidence on the left. (MMA, MP34.9.4)

construction methods and materials, limited internal watertight subdivison, lack of duplication of crucial power systems and the like) and therefore were not proper warships. When war finally broke out in September 1939 the RCN planned to build Halcyon class minesweepers as its primary small warship. However, Canadian shipyards could not work to naval standards so the RCN settled on the corvette as its auxiliary vessel. About twenty-five or so were needed to supplement or replace the hodge-podge of civilian vessels conscripted or purchased into service when war was declared. These were to be jacks of all trades attached to the key defended ports along Canada's coast: patrolling, minesweeping, doing rescue work and the myriad of duties associated with protecting a major port. The Navy also hatched a scheme whereby Canadian-built corvettes could be traded to the British for destroyers. So when all was said and done, orders for sixty-four corvettes were placed in Canada in early 1940. When the scheme to trade some of these for destroyers collapsed, ten of the contracts were transferred to the British, leaving fifty-four corvettes of the 1940-41 construction program – among them *Sackville* – on order for the RCN.

By early 1941 *Sackville*'s hull was completely plated in, while *Amherst* was nearly ready for launching. (MMA, MP34.9.5)

Design

The first Canadian corvettes differed in some important ways from the original British design, but the basics remained. The 205 foot hull was divided into three basic parts. The forward third of the ship was given over to crew accommodation. The seamen's mess was on the upper deck below the main gun. All the rest of the accommodation was on the lower deck. Directly below the seamen's mess were the stokers' quarters, aft of which lay the original Chief's and Petty Officer's mess followed by the wardroom and officers' cabins. The entire midships section (about half the length of the ship) was taken up by machinery. Two fire-tube boilers were fitted in separate boiler rooms in the middle of the ship, astern of which lay the engine room with its four-cylinder triple-expansion engine. All of these machinery spaces were open from the keel plate to the top of the engine and boiler room casings. In the engine room it was – and is – possible to see the very bottom of the ship and then look up at the sky through the skylight. Either side of the boiler rooms lay the fuel storage tanks, but the engine room remained separated from the sea only by the thin steel plates of the outer hull. Aft of the engine room lay another small messdeck where the Chiefs and POs found a new home, the tiller flats, and storage. The design was well suited to whale catching – but then whales do not shoot back.

The machinery of these early corvettes was of an old, simple and long familiar type, cheap and easy to build, easy to operate and easy to maintain. Each of the two Scotch marine boilers was basically a round drum which held a large volume of water. It was heated by 'shooting' fire through it from the three combustion chambers at the base of the boiler. Heat from the fire snaked through the larger drum in a series of pipes until it was finally vented out the funnel. These 'fire tube' boilers were large and took a long time to raise steam. Apart from the ease of construction and operation, the advantage of fire-tube boilers was that they held a large reserve of steam which was useful when a 'sprint' was needed to catch a whale – or a submarine. But the large boilers also limited the size of the fuel storage tanks on either side of the early corvettes and kept their range to about 3500 miles at twelve knots. Their maximum speed was sixteen knots.

The four-cylinder, double-acting, vertical triple-expansion engines were old technology, too, even by Second World War standards. Steam came from the boilers at up to 200psi pressure into the high pressure cylinder, then was bled off to the intermediate pressure cylinder before being sent to both of the two large low pressure cylinders at either end of the engine (hence 'triple' expansion with four cylinders). The exhaust steam was then routed into the condenser and from there back into the boilers via a condenser tank. *Sackville*'s engine, and those of her sisters, generated 2750 horsepower. The design was remarkably durable: *Sackville*'s engines soldiered on for forty years.

The only important difference in the hull between Canadian corvettes and their British cousins was the shape of the stern. The RCN decided from the outset that its first fifty-four corvettes would have to serve as auxiliary minesweepers and therefore needed wider quarterdecks to handle the equipment. Instead of the ducktail shaped stern of British corvettes (and those first ten built in Canada for the Royal Navy), Canadian corvettes were finished with a broad, somewhat square stern. This allowed the fairleads for 'sweep wire to be fitted directly astern, while leaving room for the depth charge rails and chutes.

Canadians also repositioned the galley. The British design had the galley in the aft portion of the engine room casing, which made for a long and treacherous walk along an open

Draughtsmen and engineers at Morton's yard, Montreal, with a half-model of the hull of the early corvettes, June 1942. The model well illustrates the lines of *Sackville* and the other short foc'sle corvettes built for the RCN. (NAC, PA 196835)

deck with food. The Canadians moved it forward, jamming it in just above number one boiler room. This brought the galley much closer to the crew spaces, and it also provided more room aft to fit the minesweeping winch. All of these features remain visible on *Sackville* today, and the space on her quarterdeck for working winches helps account for her long-term survival after the war.

Armament and Equipment

The only other outward difference between Canadian and British corvettes was in the placement of the after gun. In British practice (and again in those first ten corvettes built in Canada for the RN), the aft gun tub was placed between the funnel and the engine room skylight. Since the earliest plans still called for a

Sackville fitting out in the summer of 1941, with the last of the Saint John-built corvettes, *Moncton*, well advanced on the slip. (NAC PA 122192)

Amherst nearly completed. She is identical to *Sackville* in all respects except for the mainmast aft of the funnel. Note the position of the after gun tub, and the minesweeping winch and heavy davits on the quarterdeck. (NAC, PA 143933)

mainmast on the after end of the engine room casing, it was not possible to fire the gun directly astern without shooting off the mast! The RCN solved this problem by moving the after gun position to the after end of the engine room casing. This remained a distinguishing feature of Canadian corvettes throughout the war.

Like their British counterparts Canadian corvettes mounted a four-inch Mark IX breech-loading gun of First World War vintage as their main armament. The gun fired a thirty-one-pound shell a maximum of 12,000 yards with some accuracy. Its greatest value seemed to lie in forcing the submarine down so it could be pounded by depth charges, which composed a corvette's principal armament. *Sackville* and her sisters eventually carried about 100 depth charges each filled with 300 pounds of explosives (initially TNT but by mid-war replaced by much more powerful torpex or minol). These charges were either dropped from the rails on the quarterdeck, or thrown from mortars mounted on the open deck along either side of the engine room casing. The very first corvettes carried only one thrower on each side, but by the time *Sackville* fitted out in 1941 they were all equipped with two Mk II throwers per side. One thrower per side and one of the rails aft handled specially weighted 'heavy' charges designed to sink faster. By mixing the weight of the charges an attacking ship could create a three-dimensional explosion around the submarine and hope to crush its hull with the shock wave.

Important differences in weapons and electronics existed between British and Canadian corvettes. The British mounted a 2-pounder gun in the aft position and fitted .50 calibre twin machine guns (soon upgraded to 20mm) on the bridge wings as secondary armament. Canada lacked the guns and had to settle for a mix of .50 machine guns and Lewis .303 guns in secondary positions. The twin .50 guns were of some value against both aircraft and submarines on the surface, but the Lewis guns represented little threat to any enemy ship or aircraft. When *Sackville* fitted out in 1941 even the .50 machine guns were unavailable, so she went to war with twin Lewis gun mountings on the bridge and two twin Lewis mountings in the after gun position.

Canadian wireless and radio equipment and refrigeration were superior to British equipment, but early Canadian corvettes typically lacked good visual signalling equipment on the bridge. Large signal projectors (searchlights fitted with controllable shutters) were initially unavailable and *Sackville* remained limited to one large projector through much of her early wartime career. Other shortfalls were more serious. Perhaps the most important shortfall in early Canadian corvettes was the lack of a gyro compass system. Those that were available went to minesweepers, so those corvette jacks of all trades had to make do with a single magnetic compass in a standard Admiralty binnacle in the charthouse atop the bridge.

Dependence on a single magnetic compass was a serious

handicap for Canada's early corvettes. Precise navigation was difficult, and the magnetic compass was prone to error if the magnetism of the ship was altered unwittingly – from the pounding of the ship at sea, guns firing or depth charges exploding. The magnetic compass also meant that the ships had to fit the most basic of asdic (now known by the American term 'sonar'), the British type 123A: a rudimentary set which could not be easily upgraded as the war went on. The combination of a primitive asdic and a magnetic compass meant that most Canadian depth charge attacks were educated guesswork. As one senior British officer commented in 1943, "The problems of a corvette captain when attempting an accurate attack with a swinging magnetic compass are wellnigh insoluble." Fitting the magnetic compass also meant that the early corvettes lacked the lower-power electrical system required to run a gyro compass system. As a result, when weapons and sensors needed to be modernized in the middle of the war, the requirement to rewire Canadian corvettes seriously delayed their modernization.

The New War at Sea

The shortcomings in some equipment on early Canadian corvettes was known to senior RCN officers, but they planned to use the ships as auxiliary vessels in a multitude of roles. No one

Sackville at anchor trials off Saint John in late 1941, flying the Red Ensign and therefore still in the builder's hands. The flag at the top of her foremast was for a Victory Loan campaign. (DND O-53-1)

expected that they would engage in prolonged ocean voyages and battle their way through waiting packs of submarines. As late as the early months of 1941 the RCN still intended to deploy its corvettes in small groups of five at the various defended ports of Canada. These groups would sweep mines, patrol, and search for any submerged intruders. The thirty or so men aboard the ships would have enough fresh food for two or three days at sea, and then they could slip back in for more. Some corvettes, however, would also be assigned to the RN for service in the eastern Atlantic. Just what the Canadians thought these ships would do there remains a mystery.

It is a measure of the modest expectations the Navy had of its corvettes that it decided to name them largely for small towns across Canada, like Sackville, New Brunswick. Unfortunately, not all towns and villages on the list made the final selection. To avoid confusion with other warships in the British Commonwealth, the corvette destined to honour Churchill, Manitoba, became *Moose Jaw*, and "Jasper" was renamed *Kamloops*. By 1941 larger communities, like *Halifax*

The original bridge design was tiny and soon crowded by men and equipment, as evidenced by this shot of *Moncton* during her work-ups in May 1942. Note the twin .50" machine guns and the 20" signal projector. (NAC, PA-191631)

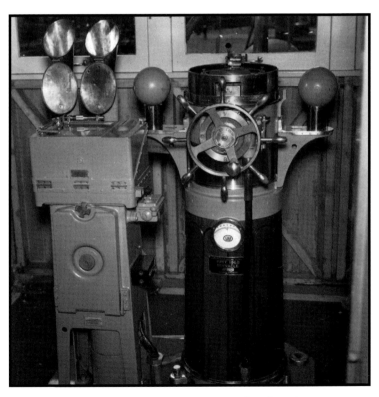

The nerve center of the early corvettes: a single Admiralty pattern 1960 magnetic compass binnacle, with a handwheel mounted to control the direction of the ship's asdic transducer, and a simple chemical recorder on the left with earphone jacks on the stand for the asdic operator and anti-submarine officer. (NAC PA 136247)

In 1940 the Navy ordered a second six-ship program of short foc'sle corvettes which were outwardly similar to the original program but which incorporated a few modifications. The most important change, water-tube boilers, is not evident in this fine shot of *Dundas*, but her lack of minesweeping gear and her bridge wing extensions are apparent. (DND E-2682)

and *Vancouver* were on the list, evidence of the corvette's growing importance.

In fact, by early 1941 corvettes were deeply engaged in the battle against German U-boats. The German forces' occupation of France and Norway in 1940 gave them advanced bases from which they could send even small U-boats to attack shipping in the depths of the Atlantic. The U-boats also adopted new tactics for both locating and attacking convoys on the high seas. Location was achieved by deploying the U-boats in long patrol lines across the convoy lanes. Once a convoy was intercepted, this "Wolf Pack" (directed by shore staff using High Frequency wireless) collapsed in on it, attacking on the surface at night, trimming down until only the sub's conning tower was above water and then running into the convoy like a motor torpedo boat.

The new German methods caught the British by surprise. It was necessary to expand the convoy system rapidly and in 1940 the British began to use their burgeoning corvette fleet for ocean escort of convoys. Through that year corvettes pushed further and further westward into the Atlantic in response to U-boat attacks. By 1941 the British were scrambling to complete an escorted convoy system across the north Atlantic.

The original corvette design, with its short foc'sle, limited crew spaces and simple equipment was not suited to this new kind of war. The British therefore immediately began to rebuild their corvettes, extending the foc'sle to improve living quarters and upgrading weapons and sensors. They also began construction of new types of corvettes with improved hull forms and better bridges.

The burgeoning fleet of Canadian corvettes in the spring of 1941 allowed the RCN to complete the trans-Atlantic convoy sys-

HMCS SACKVILLE 1941-1985

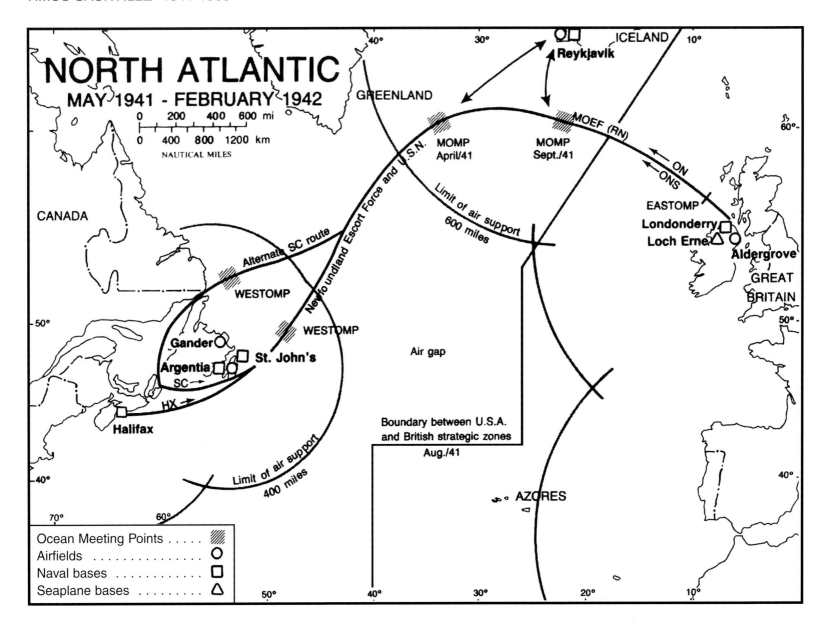

tem. Clearly, the original 1940-41 program, which was just coming to fruition, with its short foc'sles, cramped accommodation, outdated bridges and electronics, and minesweeping gear was not designed for this new role. Nonetheless, the program was too far advanced to abandon or to delay ships to fit the new modifications. The Navy chose to commit what it had to the new operations, while beginning its own program of more modern corvette designs. It was, in the end, a sensible choice since no one in Canada could have anticipated the kind of war which lay ahead.

Thus, as *Sackville* slipped into Courtenay Bay on that rainy mid-May day in 1941, senior officers were already contemplating sending the corvette fleet to Newfoundland to close the gap in trans-Atlantic convoy escort. By the end of May plans were finalized and in June the Newfoundland Escort Force (NEF) came into being. This was the start of a mid-Atlantic war that would last for another four years and the beginning of a campaign in which *Sackville* would spend all of her operational career and win her greatest distinction.

NEF took charge of the convoys between the limits of local Canadian escorts on the Grand Banks and the limits of British escorts in the east. This meant that they had to protect convoys across the most treacherous stretch of the Atlantic: the so-called Black Pit. Until 1943 that area of the North Atlantic lay beyond the range of land-based aircraft and it was far removed from the possibility of rapid reinforcement by other warships. It was, therefore, the area where the Wolf Packs hunted with impunity.

NEF had a brutal baptism of fire by any measure. Naval officers knew that the Canadian corvettes at the time were unsuited to battle Wolf Packs in the depths of the cold North Atlantic, but they understood that the actual convoy escort was the very last line of defence. The key to protecting shipping in the Atlantic lay in organizing it into convoys and using naval intelligence – including the famous Ultra signals intelligence – to keep convoys well clear of U-boats. For the most part it worked, especially when the enemy's signals could be decoded. In fact, the Allies began to read German U-boat signals with some regularity in June 1941, just as NEF was forming, so there was some expectation that those hurriedly assembled Canadian escort groups would not have to fight.

Through the summer of 1941 that was true. And there was even less expectation that NEF would have to carry the burden of the mid-ocean war when the United States – still officially neutral – began to escort convoys in the same area in mid-September. In fact, the RCN and USN shared the burden in the air gap. Through the fall of 1941 the faster American force of destroyers took charge of the fast convoys, while the RCN groups composed largely of corvettes protected the slow convoys. It was a logical division of labour, but it meant that Canadian escorted convoys were particularly vulnerable to attack.

Ultra ought to have secured those slow Canadian convoys from heavy attack during 1941. Ironically, however, the Allies were so successful at avoiding the enemy's patrol lines that the Germans, in frustration, simply dispersed their U-boats in the mid-ocean in late summer. That made it harder for naval intelligence to track them and the Canadian escorted convoy SC 42 stumbled into a swarm of U-boats in September. The battle for SC 42, in which sixteen ships were lost between Greenland and Iceland in a running battle, revealed many of the limitations of the current fleet. But the RCN also recorded its first known U-boat kill, when the corvettes *Chambly* and *Moose Jaw* sank *U501* as they closed the convoy to assist the beleaguered escort. Other battles followed in the fall of 1941, as NEF came under intense German pressure. More escorts were assigned to Newfoundland to bolster the groups, while reinforcement from British and American forces augmented the Canadian effort.

The Difficult Road to War

While the bulk of the first corvette building program hurried off to war in the North Atlantic, *Sackville* completed her fitting out in Saint John. By October she was ready for sea, and spent the next few weeks undergoing trials. Full power trials were held in December, to put her engines to the ultimate test and to record her best speed – 16.5 knots.

It's possible that *Sackville* might have languished in the builders' hands longer still. But on 7 December 1941 the Japanese attacked the Americans and the Allies in the Pacific Ocean. What had been a European war now became a world war and escort ships were desperately needed. By the end of December, with all the bugs worked out and all the basic equipment of a ship – if not yet a warship – fitted and tested, it was time for *Sackville* to join the fleet. The first drafts of her crew came aboard at 08:00 on the 30th, followed at 11:25 by Captain J.E.W. Oland, husband of the ship's sponsor and commander of the port, and a party of dignitaries. At 11:30 on 30 December 1941 Oland commissioned *Sackville* into the navy. It had taken nineteen months and three days from the signing of the contract to the ship's commissioning. This was twice the average for Canadian corvettes and earned *Sackville* the dubious distinction of being the second slowest corvette to complete (only HMS *Balsam* took longer, by nine days).

Even so, on 30 December 1941 *Sackville* was still not ready for war. Her first captain, Lt. W.R. Kirkland, RCNR, did not join until 2 January 1942, and then it took nearly ten days in Saint John to finish up a few things and become familiar with the new ship. Finally, at 12:05 on 11 January *Sackville* slipped her moor-

The underwater lines, stern and rudder of the original corvette design show well in this shot of *Kamsack* on the marine railway, Sydney, NS, about 1942. (DND CN 4042)

ings and set off for Halifax. Over the next three months her crew laboured, without much success, to prepare for war. The final outcome was not a happy one and *Sackville*'s long road to war took even longer than anyone could have imagined. Fortunately, though, the misfortunes of her work-up period conspired to put the right crew on her at the right time.

Sackville spent late January training in Halifax and doing sea work-up exercises in St. Margaret's Bay. A fisheries patrol near Sable Island followed in the first week of February, and then on the 13th she sailed on her first escort duty with convoy HX 175. Through the rest of February and March training exercises were mixed with work on the ship, including fitting of final bits of equipment (like the main gun shield), calibration of wireless sets, repairing minor damage from a ramming by the tug *Pugwash* and the like.

In late February *Sackville* tried her wings by escorting a convoy as far north as Newfoundland, where she paid her first visit to St. John's. On the way home she was diverted from her convoy to rescue seamen from the Greek ship *Lily*, sunk by *U587* from convoy ONS 68 east of Sable Island. It was then that rising tension between *Sackville*'s captain and his officers came to a head. The captain, it appears, drank while at sea and was abusive to his officers. The first lieutenant later complained that the captain was not able to discharge his duties and the ship had often been left in the hands of very junior officers with no experience whatever. The crisis came following the rescue of *Lily*'s survivors, when *Sackville* was unable to relocate the convoy. The first lieutenant took the extraordinary step of confining the captain to his cabin and assuming command of the ship. Unfortunately, neither the first lieutenant or anyone else knew how to navigate and it was necessary to bring the captain back to the bridge periodically in order to get the ship home.

By the end of March 1941 *Sackville* ought to have been ready for assignment. But clearly she was not. According to the Captain (Destroyers) Halifax, the officer responsible for the readiness and fighting efficiency of the escort fleet, *Sackville* made little progress during her work-ups. In the end, Kirkland was discharged from the navy as "unsuitable". Such problems were not unusual in the early days of expansion: the alcoholic first captain of *Amherst* had to be removed at gunpoint. It was more unusual that *Sackville*'s crew was dispersed throughout the fleet. That action probably stemmed from the fact that the entire crew of *Baddeck* was available to go aboard at a moment's notice. They took charge of *Sackville* on 6 April 1942.

Ready at Last

Baddeck's crew had already spent a winter on the North Atlantic run. With them came Lt Alan Easton, RCNR, a gifted merchant seaman and leader, under whom *Sackville* would enjoy perhaps her greatest days. In his classic memoir *50 North* Easton titled his chapter on *Baddeck* "The Knave". She had been plagued with engine troubles since commissioning and by the spring of 1942 was languishing – with Easton and her experienced crew – alongside at Halifax. It was a wonderfully simple idea to switch *Baddeck*'s crew over to *Sackville* and let them get on with the war.

And they did, with enthusiasm and energy under one of the best escort commanders of the war. In April "the Queen", as Easton called her in his memoir, received a new white and pastel camouflage paint scheme, her first radar set (the Canadian built SW1C) and then spent the next few weeks working-up in Halifax and St. Margaret's Bay: firing guns and depth charges, training asdic operators on the British submarine *P-514* and undergoing a final inspection.

By the time *Sackville* was ready for operational assignment the war at sea had pushed westward. American entry into the war opened vast new areas for German attack along the eastern seaboard of North America, and heavy losses were already being experienced by shipping off the US coast. The RCN began to escort convoys to Boston in March and by May was planning to extend convoys further south, to run its own oil convoys to the West Indies and to establish convoy systems in Canadian waters. *Sackville* might well have been assigned to any of these operations, but she was not. On 15 May 1942 – one year to the day after she first hit the water – *Sackville* slipped her lines in Halifax and shaped a course for St. John's to join in the mid-Atlantic battle. The Queen was finally going to war.

Short foc'sle corvettes were not designed for trans-Atlantic escort work, and were wet under most sea conditions. In this case *Battleford* buries her bow into a wave while escorting a convoy in November 1941. (DND M803-R)

CHAPTER 2
BATTLING THE WOLF PACKS

By the time *Sackville* reached St. John's on 17 May 1942 the war in the North Atlantic had changed. The action was now concentrated further south, off the US east coast, in the Gulf of Mexico and the Caribbean. The mid-Atlantic was comparatively quiet in early 1942, and that had allowed a major reorganization of forces there in February in order to shift escorts to other theatres of war. The relay point in Iceland was abandoned, and the Newfoundland Escort Force was absorbed into an Allied "Mid-Ocean Escort Force" (MOEF, see map page 24) that ran directly between Newfoundland and Northern Britain. The Canadian component of the new MOEF, designated appropriately enough as 'C' groups, now used Londonderry, Northern Ireland, as their eastern terminus.

The new escort arrangements along the North Atlantic run in the spring of 1942 were a calculated risk. Although the distance from St. John's to Londonderry was well within the range of corvettes, their need to zig-zag around the convoys and pursue U-boats meant that they could just manage the crossing without refuelling. It was a tight squeeze which also limited routing options for the convoys and put them at greater risk of interception. When the U-boats discovered this, and when their success elsewhere declined, *Sackville* and her consorts would have their fill of action in 1942.

Whether by design or simple good fortune, *Sackville* was assigned to MOEF group C.3 which was built around the two powerful destroyers *Saguenay* and *Skeena*. When that group's corvette component went to refit en masse in May 1942 they were replaced by *Galt*, *Sackville* and *Wetaskiwin*. In preparation for their new assignments the three corvettes were given a brief period of intense training in Newfoundland by Cdr J.D. "Chummy" Prentice, RCN in *Chambly*. Prentice had led the training patrol which sank *U501* the previous September during the battle for SC 42, and he had renewed his training scheme for corvettes in the spring of 1942.

The Mk IX 4-inch breech-loading gun was the main armament of all corvettes prior to 1944. This one is on *Arvida* in late 1943 or early 1944: the rails mounted on either side of the gunshield for illumination rockets were a 1943 innovation. (NAC PA 184185)

23

HMCS SACKVILLE 1941-1985

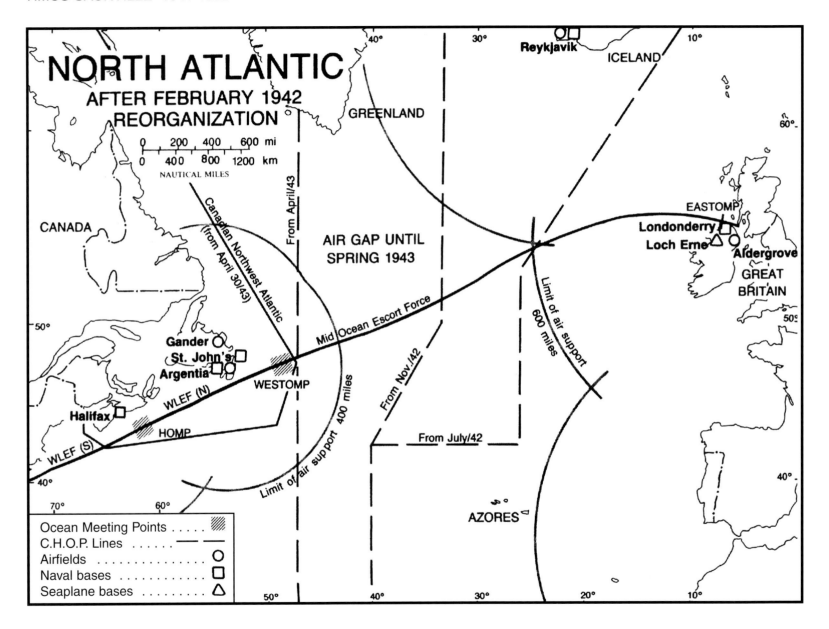

The "Revised Patrol Vessel" modification to the basic corvette design increased the sheer and flare of the bow, raised the bridge and extended the foc'sle break aft to the funnel. All of these improvements show in this splendid shot of *Port Arthur* from early 1942 (compare with the shot of *Dundas* in chapter 1). (Macpherson)

Prentice believed in realistic training and this nearly led to an untimely end for *Sackville*. She was busy attacking an asdic contact (in fact a school of fish) in a Grand Banks fog when Prentice – steering to the sound of exploding charges – brought *Chambly* onto the scene at something approaching full speed. As *Chambly* careened through a sea littered by stunned fish Easton brought *Sackville* back to the area of the attack, bursting out of the fog on a collision course. When both corvettes took avoiding action in the same direction it looked like a collision was inevitable. Each Cox'n pulled on his wheel for all he was worth and when the two ships steadied on parallel courses they were "close enough for each ship's respective [depth charge] thrower parties to almost shake one another's hands!"

Greater danger waited offshore, however, and it was fortunate that *Sackville* joined C.3, the best RCN group in the mid-ocean during 1942. It enjoyed steady leadership, steady composition and good captains – and a healthy measure of luck. All of this contributed to the development of a unique esprit de corps, distinguished by the painting of a barber pole band around the ship's funnels and the nickname "The Barber Pole Group". (When the group went to refit en mass in early 1943 the barber pole logo was pinched by C.5!)

Sackville sailed on her first operation as part of the Barber Pole group on 26 May, when C.3 departed St. John's to join the fast eastbound convoy HX 191. The passage was uneventful and the group arrived in Londonderry on 5 June. After a short layover C.3 headed westward with the slow convoy ONS 104 and was back in Londonderry on 15 July after escorting a slow eastbound convoy, SC 90. So far nothing of note had happened, but that was soon to change.

Into Battle with C.3

On 25 July 1942 C.3, composed of the destroyers *Saguenay* and *Skeena* and the corvettes *Galt*, *Sackville*, *Wetaskiwin* and now *Louisburg*, left Londonderry to escort ON 115 to Canada. By now North Atlantic convoys were once again under attack.

Twelve days earlier the Germans had deployed a pack of U-boats, appropriately code-named "Wolf", 600 miles west of Ireland. Since the Allies were suffering a gap in their reading of German radio traffic during most of 1942 it was hard for naval intelligence to pin down precisely where the subs were. On 22 July Wolf intercepted the westbound convoy ON 113 escorted by C.2, and by the 24th a battle had developed. The destroyer *St. Croix* managed to sink *U90* on the first day, but the U-boats – virtually all on their first Atlantic patrol – sank three ships from the convoy before it reached Newfoundland waters.

The battle for ON 113 pulled the U-boats to the western side of the Atlantic, leaving ON 114 a clear passage and clearing most of the eastern and mid-Atlantic for ON 115. However, a

Checking ditty-bags in the seamen's mess of *Battleford*, November 1941: not yet crowded and everyone nicely turned-out in regulation clothing. (NAC PA 184186)

Mk II depth charge thrower on *Owen Sound* in 1944. All Canadian corvettes were fitted throughout the war with the Mk II, which fired both the charge and its cradle into the sea. (DND 0-14-13)

Few corvettes survived a torpedo hit long enough for the crew to get off, let alone have someone take a photo. But *Levis* was struck well forward, in the messdecks, killing seventeen of her crew: she then took hours to sink. (DND O-326-1)

lucky sighting of the convoy by a U-boat en route to North America, just as it left the range of British aircover on 29 July (see map p. 24) resulted in a scramble to concentrate on the convoy. This proved largely ineffective as ON 115 and C.3 passed through the mid-ocean without sustained pressure from U-boats. Nonetheless, *Skeena* and *Wetaskiwin* found and sank one of these shadowers, *U558*, on the 31st in a long, deliberate asdic hunt. Short of fuel and now well astern of the convoy, these two escorts then went straight to St. John's. *Saguenay*, also short of fuel, followed them a few hours later. With that the strength of C.3 was reduced to the three corvettes.

ON 115 was nearly home, in any event. It would soon be within range of land-based air support from Newfoundland. Unfortunately, a large concentration of U-boats lay in wait ahead, including group Pirat. Further, U-boats always found the fog-shrouded waters of the Grand Banks a good place to hunt since the fog screened them from the poor radars of Newfoundland-based aircraft. The battle for ON 115 ought to have been over by 31 July, but with C.3 now reduced to *Galt*, *Sackville* and *Louisburg* the second round was about to begin.

The Germans had placed group Pirat across ON 115's path, and it was soon joined by a few stragglers from group Wolf. They regained contact with the convoy as it reached the Grand Banks on 2 August. On the same day *Agassiz* joined from Newfoundland, bringing the escort strength up to four corvettes. She made first contact with a U-boat of Pirat, and with *Galt* set off in pursuit while the convoy altered course to close faster with two British destroyers also en route to bolster the escort. By nightfall on 2 August all six escorts were in the screen around ON 115. *Sackville* pursued and attacked a false contact around

HMCS SACKVILLE 1941-1985

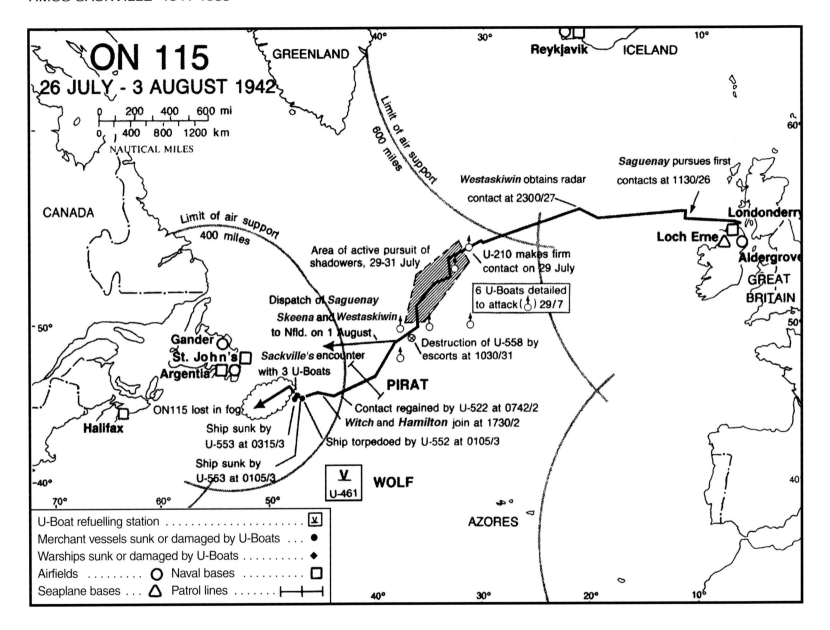

Sackville about the time of the battle for ON 115 in July-August 1942: at war only two months and already a rust-streaked veteran. Note the .303" Lewis gun and shield in the aft gun position: there is another immediately behind it on the portside. The antenna at her masthead is for the SW2C radar. (Macpherson)

midnight, and then a few minutes later two ships in the convoy were struck by torpedoes.

As soon as it was clear that the convoy was under attack the escort conducted operation "Raspberry", a pre-arranged illumination by firing starshells followed by a search astern of the convoy. As *Sackville* came back to ON 115 she found *Agassiz* and the British destroyer *Hamilton* taking survivors off the two stricken ships. Easton decided to screen the rescue work and put *Sackville* into a pattern around the scene. Forty minutes later *Sackville*'s SW2C radar set obtained a contact

The SW2C set operated on a 1.5 meter wavelength and produced only a blip on an "A" scan radar screen (much like a modern heart monitor). Easton, therefore, only knew that something was out there, but he had no idea of what or even how big. Even icebergs were a problem for early radars. *Sackville* had to approach the contact cautiously until it was clear what the target was. "Submarine, I think," one of Easton's officers reported hesitantly when the target finally came into view. "Looks more like a fishing trawler to me!" Easton answered, trying to sort out the silhouette in the dim light. It turned out to be *U43*.

"Full ahead!" Easton ordered, followed immediately by an order to his first lieutenant to "Fire!" *Sackville*'s starshell looped out and burst behind the U-boat, leaving no doubt whatever of its identity. Easton set course to ram while the four-inch main gun

29

SS *Belgian Soldier*, torpedoed, abandoned and adrift astern of ON 115 as seen from *Sackville*'s foc'sle in the early hours of 3 August 1942. (DND PMR 84-0075)

blasted away. Before *Sackville* could ram *U43* it disappeared in a crash dive. "Set pattern A!" Easton now ordered, preparing his depth charges for a shallow setting as the corvette rode in over the swirl. *Sackville*'s depth charge crews managed to get five away, boiling the ocean and sending towering plumes of spray skyward.

Easton could see little from the bridge and fretted on his failure to hit the sub with gunfire, to ram it and now, he believed, to damage it with depth charges. He swung *Sackville* back in a tight turn to repeat the dosage when the officer in charge of the depth charges arrived on the bridge in a breathless state. "It was the finest thing you ever saw, sir!" he exclaimed. "We are wasting ammunition now!" As Easton recalled in *50 North*,

The depth charge from the starboard thrower sank fifty feet and then exploded, as did the others. It must have touched the U-boat's after deck as it went off, for a moment later the bow of the U-boat broke the surface a few feet astern [of *Sackville*]. She rose up out of the water to an angle of about forty degrees exposing one-third of her long slender hull ... As she hung for an instant poised in this precarious position, a depth charge which had been dropped over the stern rail exploded immediately beneath her and she disappeared in a huge column of water.

"She'll never surface again, sir" – but she did. *U43* made port with heavy damage. At the time *Sackville* was accredited with a "probable" U-boat kill.

However, the encounter with *U43* was just the beginning. Efforts to relocate *U43* on asdic failed and a few hours later heavy fog enveloped *Sackville*. Shortly afterwards her radar gained another contact. Once again Easton had to approach with great caution due to the fog and the limits of classification provided by his primitive radar. When the U-boat was finally sighted it lay slightly on *Sackville*'s beam and the corvette – a nimble craft by any account – could not turn fast enough to ram. Easton did get close enough to hear "a colossal sound of escaping air, as though the top of a high pressure chamber had been blown off" and then "the U-boat went down like a stone." No asdic contact was ever established and Easton set off in frustration to rejoin ON 115.

While *Sackville* was chasing U-boats another sub penetrated the screen around ON 115 and torpedoed the SS *Belgian Soldier*. The crippled ship drifted astern of the convoy and was found by *Sackville*, which had, in the meantime, taken up screening *Agassiz* as she towed the stricken steamer *G.S. Waldon*. *Sackville* sent a boarding party onto *Belgian Soldier*, removing her secret books, a sextant (which remains with the ship to this day) and one crewman who had not yet abandoned ship. She then returned to screening *Agassiz* and her charge.

Sackville remained with *Agassiz* and *G.S. Waldon* throughout the morning of 3 August, as the three ships groped their way through heavy fog. At about noon *Sackville*'s asdic operator obtained a contact by hydrophone (using the asdic as a passive listening device) on what sounded like diesel engines. The radar, which was in one of its periodic rest cycles, was flashed up and used to close the target, which soon revealed itself as another U-boat. It was on a slightly closing but almost parallel course.

The U-boat was *U552* under Erich Topp, who recorded the incident after the war in his memoirs. Topp had taken advantage of the fog to lay on the surface and extract his last torpedo from deck storage and bring it below: a very hazardous move. Luckily for Topp the work was just completed when the shout "Alarm! Destroyer!" came down from the conning tower. Topp wanted to know more, but as he recalled "This was not the time for details." His officers later reported that *Sackville* came at *U552* "out of the fog like a gigantic wall ... and opened fire with every barrel ...."

Easton's view was somewhat different. His approach was cautious because he once again had no idea what he was up against. "We were now placed in the most precarious of all positions," Easton wrote later, "on a collision bearing with an invisible vessel who might be a friend: I did not know. Nor did I know her course precisely, but we stood a very good chance of ramming or being rammed ...." When the submarine finally emerged from the fog she was about to cross *Sackville*'s path a little more than 100 yards away.

"Hard aport. Full ahead. Open Fire!" Easton ordered, but the range was already too close and the U-boat was inside the corvette's turning circle. *Sackville*'s four-inch main gun would not depress enough to hit it while the bullets from the machine guns simply bounced off the hard pressure hull. Easton desperately wanted to ram the U-boat. He closed the distance, eventually down to less than 200 feet, but he could not bring *Sackville*'s bows to bear. There was only one thing left to do.

"At this moment," Topp wrote, "there is a powerful crash. The boat heaves, the lights go out." Unable to ram and too close for his gun to depress, Easton allowed *Sackville* to roll fully on her turn so that the four-inch gun would finally bear on the U-boat. One shell slammed home into *U552*'s conning tower, striking with a flash and leaving a trailing plume of smoke. It opened what Easton called "a gaping hole at the base of the U-boat's conning tower." It was a fine shot, but the damage looked far worse than it really was. Most of what the shell blew away was outer casing. A diesel ventilation pipe was ripped open and one crewman wounded. Topp managed to get the U-boat down, hastened in part by water pouring in through the damaged vent. Then the depth charges followed, pounding *U552*. *Agassiz* arrived to help, but neither corvette could regain contact: *Sackville*'s third U-boat in less than twelve hours had escaped.

"A report of our sinking arrived in Switzerland via American newspapers," Topp wrote, "and from there they came to Germany. Tears among friends, relatives and parents, and then great joy, joy in double measure when we appeared again from our patrol."

The battle around ON 115 effectively ended with *Sackville*'s attack on *U552* on 3 August 1942. Three ships had been torpedoed, two of which sank, while C.3 sank *U90* and *Sackville* inflicted damage on *U43* and *U552*. It was not a bad exchange. C.3 was, nonetheless, criticized for expending most of its fuel and effort too early in the battle, leaving the convoy poorly defended during the attacks on 1-3 August. Such were the perils of the new escort arrangement. The problem of short-ranged escorts was overcome during 1942-3 by the introduction of refuelling at sea for small escorts.

Two wonderful shots of *Sackville* in late 1942. She now sports a "Monkey Island" on top of the charthouse and the C.3 "Barber Pole" band around her funnel. (N. Cowell)

A rare shot of the open foc'sle of an early corvette, in this case *Shediac*, taken from about the galley door. All of this space was enclosed during the foc'sle extension. (Author's Collection)

The Need to Modernize

Sackville's action with three U-boats around ON 115 drew praise from senior officers on both sides of the Atlantic. Her inability to sink any of them was attributed largely to lack of modern radar, which would have allowed Easton to develop a sensible plan of attack in poor visibility. The latest British 10cm radar set, the type 271, would have given *Sackville*'s crew the range, course, speed and nature of target necessary to move through darkness and fog with certainty. As it turned out, her SW2C set helped to locate the U-boats but necessitated the cautious approach which gave them time to escape. As the British radar officer at Western Approaches Command in Liverpool, England, commented, "*Sackville*'s two [actually three] U-boats would have been a gift if she had been fitted with RDF [radar] type 271."

Sackville's battles in the Grand Banks fog proved to be a major stage in the RCN's struggle to secure modern radar for the fleet. Many of the fleet fit the equipment before the end of the year: *Sackville* had to wait until her refit in early 1943. In the meantime they all carried on in the mid-ocean against a mounting U-boat threat. C.3 was spared the worst. None of its convoys from August through to October were seriously threatened although others around them were heavily attacked. The only noteworthy event of those otherwise happy days was the loss of Able Seaman Gordon Cartwright overboard in heavy seas just north of Ireland on 14 October. Cartwright was standing quarterdeck lookout and simply disappeared. No trace was ever found.

Meanwhile the burden of the mid-ocean war was carried by the four C groups of the MOEF in late 1942. The worst befell group C.1 while escorting SC 107 eastward through the air gap in early November. The convoy was intercepted while still south of Newfoundland and was dogged by a large pack of U-boats all the way across. Fifteen ships went down, with no retribution extracted from the enemy. C.3 got slightly entangled in the building concentration of U-boats in the mid-ocean later in the month, when SC 109 became the subject of a major German search. At the outset it looked like a repeat of SC 107. The passage began badly when *Saguenay* was rammed by a ship while enroute to the convoy, sending depth charges tumbling into the sea which blew off her stern. On the 18th the tanker *Brilliant* was hit and *Sackville* escorted her part way back to St. John's before rejoining the convoy. Fortunately the Germans never established firm contact with the convoy. SC 109 slipped eastward without loss while *Sackville* and her sisters pushed U-boats aside and chased away shadowers. Once again C.3 steamed through the eye of the storm. Luck seems to have been a factor, but a less experienced and well-led group might not have been so fortunate.

The return crossing with ONS 152 proved to be the last for the old Barber Pole group. C.3 arrived in St. John's on 23

December 1942 after a gruelling winter passage: *Sackville* had only one day's fuel remaining. The North Atlantic was on the verge of the worst winter in the century. ONS 154, coming on behind C.3 and escorted by C.1, was clobbered by the tail-end of a hurricane force storm. It was also beset by a huge pack of U-boats. After the war it was determined that C.1 sank one of the attackers, but the subs claimed fifteen ships from the convoy. This was a hard blow for the RCN coming so soon after the tragedy of SC 107 in November. Through the latter half of 1942 the four C groups, thirty-five percent of MOEF's strength, had borne the brunt of German attacks in the Atlantic and suffered eighty percent of MOEF's shipping losses. They fought back gamely, sinking half of MOEF's score of U-boats, but Canadians lacked the equipment and ships to adequately defend their convoys and make the enemy pay for his success. *Sackville* and her sisters now desperately needed modernization in order to fight effectively in the mid-ocean.

The Naval Staff understood the need to modernize the corvettes of the first building programs, but it was nearly impossible to find the time, the dockyard space and the resources to do so. They hesitated between the need to keep ships at sea and the knowledge that newer, much improved ones were nearing completion. For these reasons the Navy moved cautiously on modernization in late 1942, planning to conduct a test case before moving on the seventy-odd ships that needed work. During the fall of 1942 *Sackville* was actually designated as the lead ship in the modernization program.

As things turned out, *Sackville* and the rest of C.3 went to refit, not rebuilding, in early 1943. After her arrival in St. John's from ONS 152 on 23 December *Sackville* stayed only long enough to refuel. By Christmas Eve she was en route to Halifax, where she arrived on Boxing Day. On 13 January 1943 she steamed the short distance to Liverpool, Nova Scotia, to begin twenty weeks of refitting.

During the four and a half months that *Sackville* refitted in Nova Scotia the Atlantic war reached its climax. The British took over responsibility for MOEF in January, sending the C groups for re-equipping and training under British direction. The RN then had the unhappy duty of escorting slow convoys through the air gap, and their convoys, too, suffered badly. But they had the resources to fix the problem. After stunning U-boat successes in February and March the British introduced "Support Groups" to provide aid for the threatened convoys as they passed through the mid-ocean. And along with the Americans they found, by the spring, enough very long range patrol aircraft to eliminate the gap in land-based air support in the mid-ocean entirely. By April the Allies – largely RN forces – were on the offensive in the North Atlantic. Losses to convoys fell dramatically, while U-boat sinkings rose sharply.

Sackville missed this climax in the Atlantic war. She lay alongside at Liverpool until late March, having her machinery stripped down and rebuilt, and her weapons serviced. Only a few structural changes were made. The bridge wings were heavily reinforced to carry 20mm secondary armament, and the bridge itself was extended forward to fit a second magnetic compass binnacle. The latter modification would allow the ship to be 'fought' by the captain from the bridge. With this new arrangement the "monkey island", built atop the charthouse prior to the battle for ON 115, was removed. *Sackville* also finally received her 2-pounder pom-pom for the after gun position. With that gun, and the 20mm oerlikons on either wing of the bridge, she now carried the kind of powerful secondary armament that would have made short work of those three submarines the previous August.

With her machinery restored and the bridge slightly improved *Sackville* steamed to Pictou in late March to have her minesweeping winch removed. At that time, or possibly during

Modernization of the ten British-owned corvettes operated by the RCN started early: this is *Trillium* having her bridge and foc'sle rebuilt in the US in early 1942. (DND PMR 83.395)

her stay in Liverpool, the distinctive gooseneck minesweeping davits on the quarterdeck were replaced by a small davit for handling depth charges. Then it was back to Halifax for docking.

It was during this latter stage of her refit that *Sackville* also received a new captain. Lt Alan Easton, RCNR, left the ship for a posting ashore prior to taking command of the new frigate *Matane*. Frigates, originally known as "Twin Screw" corvettes, represented the ultimate development of the basic corvette idea adapted for ocean escort. They had twice the propulsion machinery (hence two propellers), one hundred additional feet of hull, a bridge to naval standards and the latest in weaponry and sensors. Easton's replacement in *Sackville*, Lt. A. H. Rankin, RCNVR, assumed command on 10 April 1943.

Back to Sea with C.1

Rankin took *Sackville* through her training and work-up exercises in April, and then sailed her on 4 May for St. John's and the broad reaches of the North Atlantic once more. She was in fighting trim after nearly twenty weeks of refit, re-equipping and training, but two outstanding limitations remained. Despite the length of time she lay in dockyard hands, *Sackville* remained a short foc's'le corvette. Designed for a crew of twenty-nine all ranks, she now carried just a few short of ninety. They jammed into the forward messdecks as best they could, but the ship was terribly crowded.

The short foc's'le also meant that *Sackville* remained a very wet ship. The open welldeck between the deckhouse and the foc's'le was aptly named, since it seemed to attract water. Any sea shipped onto the foc's'le simply ran aft and cascaded over the lip of the foc's'le into the main crew thoroughfare. At action stations the welldeck hatchway to the lower mess had to be left open so that the shell hoist to the magazine could be rigged. That allowed water to pour down into the lower messes. The pervasiveness of sea water was matched by condensation formed on the deckheads and sides of the hot, crowded messes. Water dripped incessantly from overhead and in poor weather it sloshed around on the messdecks. Bringing food forward from the galley was a major accomplishment in rough weather. Short foc's'le corvettes were never intended for long passages on the broad Atlantic, nor for the scores of men who were needed by mid-war to work the increasing number of weapons and equipment. By 1943 *Sackville* and the whole Canadian corvette fleet badly needed foc's'le extensions.

Moreover, until they had their foc's'les extended, and until the ships were re-wired with a low-power system, it was not possible to modernize the navigation equipment, anti-submarine weapons and asdics of Canadian corvettes. The two projects – rebuilding and rewiring – really needed to go hand in hand. Until it was done Canadian corvettes lacked precision in their own operations and found it difficult to co-operate with more modern vessels. The RCN started the work in early 1943, taking *Edmundston* in hand as a test case. She was completed in May. By then the navy was engaged in a crash program to modernize as many of its old corvettes as possible.

Sackville missed the 1943 wave of modernization by the slenderest of margins. She would have to steam on throughout the year as best she could: better armed, with a modern radar and a skilled crew. The navy clearly had faith in her, her crew and in Rankin because *Sackville* was soon selected as one of the few Canadian ships to participate in the 1943 counter-offensive against the U-boats.

The 1943-44 corvette programs produced ships with improved bridges, pressurized boiler rooms and increased endurance. *Long Branch* illustrates many of these improvements, including the raised gun platform, higher bridge and -- by the absence of ventilators around the funnel -- her new boilers. The stubby mainmast is also a clue that she was built in Britain. (Macpherson)

CHAPTER 3
From the GNAT to Galveston

Few periods brought more profound change to the Atlantic war than the spring of 1943. During the time *Sackville* was in refit in Liverpool Nova Scotia, Allied fortunes in the mid-ocean went from near catastrophe in February and early March, to a major victory over the U-boats in April and May. In the first five months of 1943 a hundred U-boats were sunk in the North Atlantic, hounded by increasingly oppressive air cover and harried by the new "support groups" that shepherded convoys through the danger areas.

During most of 1943 RCN escorts formed the inner ring of this highly effective new system. The destruction of U-boats was usually left to aircraft or the bigger and better equipped destroyers and frigates of the RN support groups. The latter even took over good contacts picked up by the close escorts – leaving them to move on with the convoy – and hunted them until the U-boat was destroyed. The RCN lacked the destroyers and frigates needed to participate fully in this counteroffensive. Nonetheless, it did cobble together two support groups of its own, partially composed of corvettes, by mid-1943. These groups produced some of the most dramatic moments in Canadian naval history. *Sackville*, fresh from refit, re-equipped and with an experienced captain and crew helped to make that history.

Sackville's initial assignment in May 1943 was to MOEF group C.1, built around the British frigate *Itchen* with Cdr C.E. Bridgeman, RNR in command and as senior officer. The group included the Canadian destroyers *St Laurent* and *St. Croix* and the corvettes *Agassiz*, *Napanee* and *Woodstock*. C.1 was already in Britain by the time *Sackville* was ready for operations, so she

Hard and dangerous work: loading a "heavy" depth charge (the sailor in the white cap has his hand on a weight added to make the charge sink faster) onto the cradle of one of *Mayflower*'s starboard throwers. The nearness of the two throwers was characteristic of the original British design. (DND NF 338)

Shediac refuelling at sea in 1943. By then it was a routine occurence in the North Atlantic, but without high pressure hoses it was a long and laborious business. (Author's collection)

When veterans complained that corvettes were wet ships, they meant it! *Napanee* shows her keel while her entire superstructure and mast are obscured by spray. (DND CN 3295)

sailed directly to Londonderry to join them. The group then steamed westward as escort for ON 184.

Sackville's veterans would soon have noticed significant changes in the way convoy escort operations now ran. All of C.1's convoys in the spring and summer of 1943 were fast ones, since the RN now carried the burden of the slow convoys. This meant that Canadian convoys spent less time in the danger area and were better able to slip away from trouble. Between 13 May 1943, when *Sackville* and C.1 left Londonderry to escort ON 184, and mid-August when she joined a support group herself, C.1 escorted six convoys across the Atlantic without incident. All were HX or ON, and made the transit of the Atlantic in little more than two weeks – fast and easy by 1942 standards.

In the spring of 1943 each North Atlantic convoy was now joined by a support group the day before contact with U-boats was anticipated, and those additional forces stayed with the convoy until the danger had passed. ON 184, for example, was supported by group EG 6, which was built around the small USN escort carrier *Bogue*. And if that was not enough, very long range Liberator patrol aircraft now appeared routinely in the depths of the mid-Atlantic to provide additional support. Gone were the days when *Sackville* and a few other escorts would have shepherded a slow convoy through the mid-ocean without air support or reinforcement. In addition, the Allies now stopped trying to avoid U-boat concentrations. Instead, they sent convoys along the most direct route across the North Atlantic in order to force a battle which the Germans could not win. Corvettes like *Sackville* had plenty of fuel to make the more direct crossing. Nonetheless, refuelling at sea was now a regular practice, and *Sackville*'s log records topping up fuel from the escort tanker on every crossing. Thus, compared to 1942 there was not a lot for C.1 to do, other than keep the convoy in order and see it through.

Meanwhile British support groups and convoy escorts sank U-boats all around Canadian escorted convoys, which were steered well clear of trouble – even in the decisive month of May. During that month the U-boat fleet suffered terrible losses. By the end of May the Germans finally abandoned Wolf Pack operation in the North Atlantic, a defeat from which they never fully recovered.

While the close escort work turned into something of a milk run, the real battle against the U-boats was now fought by support groups. When the mid-ocean U-boat campaign was defeated at the end of May support groups switched to hunting them in their transit areas like the Bay of Biscay. Canadians were anxious to participate in this new offensive and by July had formed one support group, designated EG 5, which by August was serving in the Bay of Biscay.

Joining the Hunters of EG 9

In late July when the RCN was asked to form another support group it took the core from C.1: *Itchen*, with Bridgeman as senior officer of the new group, *St. Croix* and *Sackville*. *Chambly* and *Morden* joined from C.2. Since there were very few modernized corvettes yet available, and with most already assigned to EG 5, it was difficult to find ships for this new group. The deciding factor, in the end, seems to have been a mixture of size, equipment and speed in the larger ships, and the right people in command of the corvettes. The new support group, designated EG 9, trained in Britain in early September. On the 12th it arrived in Plymouth ready to join in the offensive in the Bay of Biscay.

The "Twin Screw Corvette" was such a vast change in design that the Chief of the Canadian Naval Staff, Percy Nelles, recommended that they be redesignated "frigates". The British agreed and that term came back into use. This is the first Canadian frigate, *Waskesiu*, in 1943. (DND DB 0167-16)

By the time EG 9 was ready for war the anti-submarine offensive in the Biscay was virtually over. The Germans had moved to protect their submarines with heavy airpower, including the introduction of a new air launched glider bomb. That new weapon was first employed in late August against the Canadian support group EG 5, another British group and their supporting destroyers patrolling off northwestern Spain. The bombs, which were remotely controlled from a large aircraft, contained enough explosives to completely shatter a corvette. Several of EG 5's corvettes suffered near misses, and in the process discovered reserves of speed they never knew they had! Larger ships with wider turning circles seemed easier to hit: one British sloop was destroyed and the Canadian destroyer *Athabaskan* severely damaged by the glider bombs.

Few shots illustrate the dangers of winter service off the Canadian east coast better than this one of *Lunenburg* in early 1943. The weight of ice build-up made the ships unstable and equipment could not be used. (DND PMR 92-102)

Fortunately, perhaps, *Sackville* and the rest of EG 9 were spared the danger from glider bombs. By the time they were ready to conduct offensive patrols the Germans had renewed the threat to the main trans-Atlantic convoys. Following the defeat of the Wolf Packs in May 1943 the Germans re-equipped their submarines in hopes of renewing the mid-ocean campaign. Heavier anti-aircraft armament would ward off Allied aircraft, and the new German Navy Acoustic homing Torpedo (GNAT as the Allies called it) would blast a way through the naval escorts and allow the U-boats into the convoys. The GNAT had an active homing mechanism that focused on the high-pitched sound of a warship's propellers. It also had a magnetic proximity fuse so that it was not necessary to actually strike a small, shallow and fast moving escort. By mid-September 1943 Wolf Pack "Luethen", composed of re-equipped U-boats, assembled west of Ireland, ready to snare the next series of westbound convoys.

Battling the GNAT

Instead of hunting U-boats in the Bay of Biscay, *Sackville* and EG 9 set off to find them in the mid-ocean. EG 9 was assigned to support two convoys which were routed so that they arrived in the danger area together, allowing for maximum use of the naval and air escort. ONS 18, the slow convoy, led the way escorted by B.3, while ON 202 escorted by C.2 closed from behind. A Canadian Liberator opened the battle on the 19th by sinking *U341*, while the first U-boats made contact with ONS 18 by midnight. Meanwhile, EG 9 approached the scene from the southwest, monitoring the action on radio.

Although ONS 18 seemed the most threatened, it was ON 202 – forty miles astern – that bore the brunt of the first night's battle. The British frigate *Lagan*, part of C.2, went out to investigate a contact and had her stern blown off by the first GNAT fired in anger. The frigate was wrecked, but was towed to port.

About the same time a U-boat entered the convoy sinking one ship and leaving another crippled and drifting. The first round therefore went to the U-boats.

On 20 September it was agreed to combine the corvettes of B.3, C.2, and EG 9 into an inner ring of protection, while the frigates and destroyers formed the strike forces. This was not easily done, since the convoys still remained some distance apart, but in practice that was the pattern followed for the next few days. That did not mean, however, that there was nothing for *Sackville* and her fellow small ships to do. In the battle of ONS 18/ON 202 there were enough U-boats for everyone.

In fact, the pace of the battle on 20 September was frantic, with U-boats popping up everywhere. *Sackville* never got a good crack at any of them, and it's fair to say that this round, too, went to the subs. Things started to go wrong for EG 9 when *St. Croix*

The smiling faces of merchant sailors, survivors of a convoy battle, crowd the engineroom casing of *Shediac*. (DND PMR 83-1247)

Sackville following her 1943 refit. Her bridge has been extended slightly forward, and the wings properly supported to carry a 20mm secondary armament. The lantern just forward of the funnel is for the type 271 radar. Illumination rocket rails have been fitted to the main gunshield and the heavy minesweeping davits are now gone from the quarterdeck. (DND CN 3557)

set off in pursuit of *U305* late in the afternoon. The sub saw the destroyer first and *U305*'s first acoustic homing torpedo blew off the ship's stern. The second detonated directly under the keel amidships, shattering *St.Croix* like an egg. *Itchen*, closing the scene to help after the first torpedo, was driven off by the explosion of a torpedo in her wake. She could not return until another escort arrived to screen the rescue work. As a result, more than a hundred survivors of *St. Croix* spent a brutally cold night fighting for their lives in leaky boats.

Meanwhile, *Itchen* was drawn away by the need to find out what had happened to the British corvette *Polyanthus* which was supposed to screen the rescue of *St. Croix*'s survivors. The little corvette had been pursuing her own contact when a single torpedo from *U952* ripped her open and she sank like a stone. *Itchen* arrived to find a lone survivor clinging for life in the cold sea.

During the night *Sackville* foiled the only serious attempt to attack the convoys. She picked up what was probably a medium frequency homing beacon on her MF direction finder. The MF/DF was intended for navigational purposes, but it could also locate the homing beacons used by U-boats to draw other submarines onto convoys. In this case *Sackville* chased and drove down *U378*. The U-boat responded by firing an acoustic homing torpedo that fortunately missed, although it shook the ship. Perhaps for that reason Rankin considered it prudent to return to the screen around the convoy.

The next morning *Itchen* found her way back to *St. Croix*'s survivors. They were now reduced in number by hypothermia, and the British frigate took about seventy men aboard before heading back to the convoys. With the loss of *St. Croix* senior naval officers already had doubts about the continued viability of EG 9.

Heavy fog shrouded the convoys on 21 September and on into the later afternoon of the 22nd, which slowed the pace of the battle. Radar-directed sweeps by the escorts kept the U-boats at bay and allowed the British destroyer *Keppel* to surprise and ram *U229* on the surface. On the 21st *Sackville*, *Chambly*, and *Morden* swept astern of the convoys to a depth of twenty miles in an attempt to catch U-boats trying to pursue, but they had no luck. Around noon on the 22nd *Sackville* attacked a submerged contact which produced some oil, but no further traces of a U-boat. When the fog finally lifted on the afternoon of the 22nd the air was filled with Liberators. The aged Swordfish biplanes from the "merchant aircraft carrier" *Empire MacAlpine* sailing with the convoy were airborne, too.

Yet even this tremendous air cover could not drive off all ten U-boats which remained in contact. The battle re-joined that

Back to war: the crew of *Sackville* pose for photos in St John's in July 1943 following her refit. The lieutenant front and center in the group of officers is A. H. Rankin, RCNVR, who has just taken over command from Alan Easton. Note the gunshield art, soon obscured by rocket rails – a "Sack Full" of U-boats. (NAC, PA 196836 and PA 196837)

night, with the surviving members of EG 9 formed ahead in an advanced screen of the two convoys – which, somewhat miraculously, had managed to come together. *Sackville* and *Chambly* guarded the portside, with *Itchen* in the centre and *Morden* alone to starboard. Just before midnight both *Itchen* and *Morden* obtained radar contacts. *Itchen*'s first drew her to port, across *Sackville*'s bow, before another carried the frigate back across the front of the convoy almost to the other side. There she was nearly rammed by *Morden*, who was in the process of depth charging her contact.

A few minutes later *Itchen* got another radar contact, switched on her 20" signal lamp and there, in the glow of the light, lay *U666*. A brief gun battle ensued in which the U-boat was fired on by *Itchen*, *Morden* and the leading ship in the convoy's third column. The battle ended abruptly with a salvo of two GNATs from the U-boat. The one directed at *Morden* exploded in her wake. The other detonated directly under *Itchen*'s keel and may have set off her magazines. She disappeared in a towering explosion. *Morden* was so close that wreckage from *Itchen* fell on the corvette's deck. No-one knows how many men survived the explosion only to perish as the convoy churned through the wreckage. In the end one crewman from *St. Croix* and two from *Itchen* were recovered, all that remained of three ships' companies.

A few minutes after *Itchen* was lost *Chambly* chased another contact far out to port. *Sackville*, in the midst of a chaotic scene, reported no contacts although U-boats swirled around her. Meanwhile, *U238* penetrated into the ranks of ON 202 and sank three ships, escaping without damage. Another ship from ONS 18 was lost before dawn. That morning *Sackville* was sent back to survey the battle scene, but she found nothing except floating debris. Aircover arrived the next day and effectively ended the battle of ONS 18/ON 202.

Back to Escort Duty with C.2

The loss of *St. Croix* and *Itchen* ended EG 9's brief existence. When the three surviving corvettes of the group arrived in St. John's on 26 September it was already agreed that *Sackville* and *Morden* would go to C.2, and *Chambly* to C.5. The only good news for the escorts from ONS 18/ON 202 was that the Canadian Navy had already designed, tested, manufactured, and delivered to St. John's a solution to the GNAT. Fifty sets of the Canadian Anti-acoustic Torpedo (CAT) gear were waiting on the wharf when the survivors of the battle arrived.

CAT was a remarkably simple device, consisting of two pipes loosely connected so that they rattled when towed through the water. Unfortunately, the RCN initially lacked conviction that its CAT worked, and for the rest of the fall of 1943 *Sackville* carried the very complex British noisemaker called FOXER. It operated on the same principle, but was heavy and difficult to handle. Without the powerful minesweeping winch and those large gooseneck davits lost in the last refit, it appears that *Sackville* seldom used her FOXER. It was replaced by light, hand launched CAT gear by the end of the year.

Re-assignment to C.2 meant a largely quiet autumn for *Sackville*, since the group was engaged in close escort duty. After being dry-docked for a day in St. John's to check for damage from near misses by GNATs, *Sackville* sailed on 1 October with C.2 to escort SC 143 to Britain. Five more crossings followed without incident in the fall of 1943. Christmas Day was celebrated early at St. John's on 18 December, and then *Sackville* and the rest of C.2 sailed on the 20th to escort HX 271, arriving in Londonderry barely eight days later. By then *Sackville* was reporting problems with her engine which limited it to about half its normal revolutions.

In fact, by December 1943 *Sackville* was just about due for another refit. The navy planned nothing special. Certainly they had no intention of modernizing *Sackville*. Of the sixty-odd short foc'sle corvettes in the fleet they had managed to modernize only twenty-six throughout 1943, and these were much delayed because of shortages in material and dockyard space. Now that the newer, long foc'sle, improved bridge and extended endurance corvettes and the new frigates were coming into service it seemed a waste to rebuild the old ships. *Sackville* was therefore slated for a routine refit at Thompson Brothers, Liverpool Nova Scotia, starting in February.

It was probably just as well that *Sackville* was due for refit since she was no longer suited to the task C.2 was about to undertake in early 1944. The group, ably led throughout late 1943 by Cdr P.W. Burnett, RN, an anti-submarine specialist and former instructor at the RN's anti-submarine training school, was to switch to support group operations in mid-January. Modernized corvettes, with their better weapons and sensors, were needed for this hunting role and *Sackville* no longer fit the bill. The concurrent requirement to go to refit allowed *Sackville* to leave C.2 with dignity.

Modernization at Last

While her old group lay at Londonderry waiting to tackle the U-boats – which they did with considerable success – *Sackville* left for Canada with C.4 as escort for ON 220 on 16 January. In the meantime the navy had decided, in the interests of crew morale, to continue with the modernization of its short foc'sle corvettes. Instead of going to Liverpool *Sackville* was found a slot in the busy work schedule of the Todd Shipyards of Galveston, Texas. They contracted to modernize the ship for $660,000.00 – almost exactly what *Sackville* had cost to build only two years before.

Sackville arrived in Halifax on 5 February 1944 after a difficult winter passage from St. John's in escort of JH 90. She lay alongside for nearly two weeks preparing for the journey south, did a brief stint of patrol duty off Halifax, returned to load more new equipment before sailing on 19 February for Texas. Her passage south was interrupted by an urgent call to find and standby the tug *Director*, which she did until relieved by *Kitchener* on the 22nd. Then it was due south, refuelling at Key West, Florida, on the 25th and arriving at Galveston on the 28th of February. Three days later *Sackville* was secured at Todd Shipyards and in dockyard hands. Most of the crew was given forty days' leave. The transformation of *Sackville* to a modernized long foc'sle corvette would take a mere eight weeks.

There was much to be done. *Sackville* was so rusty brown from her North Atlantic service that she soon became know among the dockyard workers as The Gravy Boat, a nickname that prompted at least one brawl between the crew and the locals. The controlling item of the modernization – the change which set the

The remarkably simple RCN solution to the German acoustic homing torpedo, "Canadian Anti-acoustic Torpedo" (CAT) gear, consisted of two small pipes laid out at the end of the wire yoke. The torpedo-shaped depth gauge was attached here for trials. (DREA)

Sackville in late 1943, just prior to her modernization in Texas. (R.W. Brotherhood)

outside time limit on all other smaller jobs that had to be done – was the extension of the foc'sle. This involved plating in the main deck from the break in the foc'sle aft of the main gun position to just behind the funnel. That gave *Sackville* a second messdeck in what used to be the open welldeck. New passageways ran aft along the sides of the main deckhouse, allowing for offices and storage compartments down either side. In addition, the foc'sle extension now enclosed the entrances to the galley on either side, which meant that food could be brought into the messdecks without fear that either it or those carrying it would be swamped by the sea. As a final touch, small breakwaters were now fitted to the forward edge of the main gun position, to help shed seas taken over the bow – a simple modification that ought to have been done long before. These were the kind of changes in habitability and accommodation that *Sackville* had needed from the day she first arrived at St. John's in May 1941.

While the foc'sle was being framed and extended *Sackville*'s bridge was completely rebuilt. Everything above the wheelhouse was stripped away. The old charthouse was replaced by an open bridge, slightly raised, with a good 360-degree view, while the bridge wings were rebuilt to better take the weight of the 20mm guns. The asdic equipment was modernized and relocated to a small hut built into the forward portion of the bridge. The type 271 radar remained in its little cabin just astern of the bridge, but to clear its view forward the mainmast was shifted to a new position aft of the bridge.

One of the most important changes made to *Sackville* in Galveston was largely invisible: she was rewired with a low-power electrical system. As a result she was now able to fit a gyro compass and a series of repeaters in crucial positions throughout the ship. Previously, all co-ordination of essential data had to be worked out from different sources using a single magnetic compass and the native intelligence of the officer in charge. With a gyro compass system fitted, all the key components in the ship – radar operators, asdic operators, the cox'n at the wheel and the officer of the watch – were singing from the same hymn book. They all shared the same information on direction from their own compass repeater, which was stabilized against the movement of the ship. Further, with a gyro compass it was now much easier to pass information from one ship to another, something that was crucial in co-ordinated asdic hunts for U-boats.

Fitting a gyro compass also meant that *Sackville* could now fit a modern asdic set and the latest in anti-submarine weapons. The old type 123A asdic set was stripped out and replaced by a type 127DV. This set allowed the operators to control and better monitor bearing and depth information on the target, and it required a gyro compass for proper control and accuracy. The type 127DV was also the basic set required to control the firing of the new ahead-throwing anti-submarine weapon, hedgehog.

Hedgehog represented a new concept in anti-submarine weapons. Depth charges had to be thrown astern of the ship, which meant that for a few minutes during the approach to the attack it was impossible to maintain asdic contact on the target. To place the depth charges properly the captain had to estimate the submarine's depth, position, course and speed from the information provided by the asdic. Then, using that information and the sinking rate of depth charges, he had to drop the depth charges where he estimated the sub would be by the time the charges sank to the right depth. It was all very complicated. And in the run-in to the attack position, the asdic – which was set at a fixed angle of depression – could not be pointed down to maintain contact on the target. Using depth charges then meant that the last two or three crucial minutes of an attack were conducted blind.

Hedgehog was designed to fire ahead of the ship while the target was still held in the asdic beam. The type 127DV asdic was used to control the weapon, which was fired when the right position was reached. Gyro-stabilization allowed the hedgehog to compensate for anything up to a 20-degree roll. A single hedgehog mounting fired twenty-four small bombs filled with thirty-two pounds of torpex a distance of 200 yards ahead of the ship, where they fell in an oval 120 feet wide and 140 feet long. The bombs themselves were contact fused, so they detonated only when they struck something, and they worked to a depth of 1300 feet. The whole system promised a vast improvement in lethality. Whereas the standard depth charge attack averaged about a six percent success rate, the hedgehog promised nearly thirty percent. Thus the combination of modern asdic and hedgehog was a crucial component of modernization.

Some thought was also given in 1944 to refitting corvettes with the new Mk IV depth charge throwers, which retained their carriers when fired. It was decided however, due to short supply of the new throwers, to limit their allocation to frigates and other large escorts. Thus *Sackville*'s thrower crews and those in all the other corvettes laboured with a two-stage loading process for the balance of the war.

A few other notable changes resulted from the Galveston rebuild. The SW2C Canadian-built radar had been retained because of its excellent air-warning capabilities, but a new, much smaller "X" shaped antenna was now fitted at the masthead. The designation of that radar now changed to SW2CP. The corvette's boat and life raft arrangements were also brought into line with the latest practice. When *Louisburg* was sunk by an aerial torpedo in the Mediterranean in February 1943 (the only RCN corvette lost to aerial attack) many men perished in the water when her own depth charges exploded. Survivors needed something that would float free and be easy to get into. It was decided to remove the two sixteen-foot rowboats on either side of the funnel and replace them with one twenty-seven-foot whaler and enough Carley floats (oval shaped life rafts) for one hundred and fifty percent of the ship's company. The Carley floats would allow survivors to get out of the water quickly and avoid the crushing effects of exploding charges as their ship sank beneath them. *Sackville*'s boat and raft arrangements were altered to this new standard while in Galveston.

Back to Sea with C.2

The Todd Shipyard was as good as its word, and *Sackville* began trials on her machinery on 29 April 1944. Full power trials were conducted on 4 May, fuel and stores completed, her new gyro compass swung on the 6th and the next day she sailed for Halifax. The only complaint recorded of the work done in Texas was on the hammock hooks. They were welded to the deckhead by an attractive young woman who had to stand on a stool to do the job. As *Sackville*'s crewmen passed they could not resist patting her backside. "You Canadian boys are gonna be sorry!" she warned. And they were. When the off-duty watches turned in on the first night out to sea all their hammocks crashed to the deck!

Unlike the hurried trip south, the cruise home was conducted at a leisurely pace, and it was not until 15 May that the 'new' *Sackville* was secured to jetty 5 in Halifax. On the 17th Lt Rankin took leave of the ship and command now passed to Lt A.R. Hicks, RCNVR. Hicks and *Sackville* spent the rest of May in Halifax seeing to a few final tasks, like checking the ship's magnetic signature, completing further machinery trials, and loading ammunition. On 1 June *Sackville* headed south again, this time to the RCN's new work-up base at Bermuda.

Sackville trained in the azure blue seas of Bermuda for three full weeks. Everything from towing exercises and refuelling at sea, to gunnery, asdic training on a real submarine and group exercises was practised. While she was there the Allies landed in France, the famous D-Day operation of 6 June 1944 that opened the final chapter in the war. Nineteen of *Sackville*'s Canadian counterparts were there for the operation, including eight of the first building program ships. However, apart from the terror imposed on her crew by the training officers at Bermuda, *Sackville* spent those days working up in the tropical sun. Everyone knew it could not last. In fact, the navy had decided that this veteran of two North Atlantic campaigns ought to undertake another one.

On 23 June 1944, now completely rebuilt and fully equipped for the kind of weather, sea and operational conditions she would encounter along the North Atlantic run, *Sackville* slipped through the Town Cut from St. George's, Bermuda, and set a course for St. John's once again. Waiting there for her was C.2 and, given the lateness of the season, the prospect of her first full winter on the North Atlantic convoy routes – if the war lasted that long.

Above and right; *Sackville* in her final fighting form following the Galveston refit, seen here while serving as the training ship for HMCS *King*'s. Her bridge has been completely rebuilt, mast re-sited aft, foc'sle extended, 16' boats replaced by a whaler and carley floats, 2-pounder gun aft, and more. (C.C. Love)

CHAPTER 4

The End of One War and the Start of Another

Sackville began her third campaign in the North Atlantic on 29 June 1944, when she sailed with C.2 to escort HX 297. To date she had been rather lucky. In battling the Wolf Packs in 1942 and 1943 she had seen more action than many wartime escorts. There was a belief, which lasted until records were verified after the war, that she had sunk at least one sub in the fog around ON 115 in 1942 and heavily damaged another. And she had fought and survived the deadly cat and mouse game of convoys ONS 18/ON 202 in September 1943, when new German torpedoes claimed several of her sister escorts. Over the years her refit schedule had also kept her from serving a full winter season in the Atlantic: so far at least. But as Sackville faced the prospect of her first full North Atlantic winter, everyone was betting that the war would be over by Christmas. When that happened, Sackville and all her war-worn classmates would be discarded by the navy. In June 1944 it certainly looked like her days were numbered.

The Flaw that Saved the Ship

In fact they were, but not in the way most anticipated. During the layover in Londonderry in early July Sackville was taken in hand for routine boiler cleaning and repairs. No defects were noted, other than some heavy caulking in the seams of the centre combustion chamber of number one boiler. When steam was raised again the chamber was found to be leaking. The faulty seams were repaired, tested and found to be satisfactory. Sackville sailed on boiler number two as C.2 departed on 22

(C.C. Love)

July, raising steam slowly in number one and watching the pressure. When the pressure reached 140 psi, the seams at the back end of the central furnace failed again, and Sackville was ordered back to Londonderry. More repairs followed, as did another water pressure test which proved satisfactory. But when steam was raised once more the boiler failed at 150 psi.

Extensive inspection of the combustion chamber and its materials was conducted by British engineering specialists in Londonderry. The boiler, as they observed, was little more than two years old and ought to have had much more life. In the end British engineers determined that faulty manufacturing had caused the plates of the chamber to fail along the rivetted seams. In addition to the obvious cracks, X-rays of the faulty seams revealed a spider's web of fine fractures radiating from the rivet holes. They concluded that the plates were improperly drilled,

Late war asdic was a vast improvement over the early type 123A set fitted in the first building program. This type 144 set, in *Coburg* in 1944, was similar to *Sackville*'s final suite. Compare this with the asdic equipment illustrated in chapter 1. (NAC PA 134330)

A firetube boiler from the first building program, in this case one built for *The Pas*. *Sackville*'s number-one boiler failed in the center of the three combustion chambers illustrated in this shot. (GMM 988.17.28.50)

were too deeply countersunk for the rivet heads, were poorly rivetted and too heavily caulked. All this allowed water to seep into the joints which, combined with the heat of firing and chemicals in the water, had produced cracking of the metal. The diagnosis was "caustic embrittlement" of the plates in the centre combustion chamber.

It was, the Admiralty specialists observed, a very rare defect. There was initial concern that *Sackville* had been steaming too long using highly acidic water, but a review of her water treatment history revealed sound practice. It was also suggested that the high acid level of water taken on in Texas might have contributed to the defect. But in the end it was concluded that poor manufacturing was the real cause. The only solution was the complete replacement of the combustion chamber.

The RCN ordered *Sackville* home so they could assess the problem. She sailed with ONS 248 on 11 August, and seems to have acted as plane guard for the two small aircraft carriers operating with the convoy. Tucked in behind the plodding merchant ships of column 10, *Sackville* steamed along easily on one boiler and her attentiveness was rewarded. On the 16th she plucked

Sackville's bridge as rebuilt in 1944. The bridge repeater for the new gyro compass system is fitted just ahead of the magnetic compass, with the new asdic hut at the top of the photo. The range finder to the left of the magnetic compass was added in 1945 to assist in loop laying operations. (DND HS 1355-4)

two flyers from their ditched aircraft and sent them back to their ship. *Sackville* finally secured to the corvette jetty in Halifax on the 26th. The RCN's own engineers now had a look.

Had *Sackville*'s boiler failed earlier in the war when escorts were badly needed the navy would have fixed it. But by August 1944 the end of the war was in sight. The Russians were closing in on Berlin and the German army in the west was in flight across France and Belgium. "Home by Christmas" in 1944 looked like a good bet. In any event there were now many far more modern escorts available to carry the burden. *Sackville*'s saving grace was that she was recently modernized and, apart from number one boiler, in excellent condition. As a result, on 29 August 1944 she

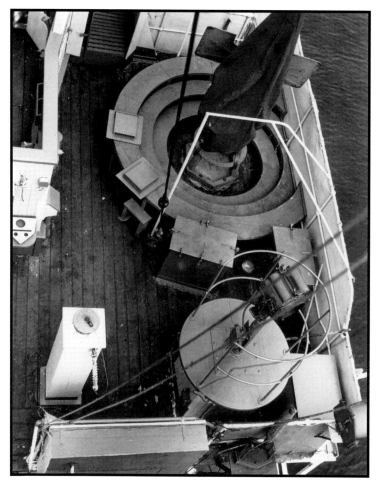

Starboard bridge wing as rebuilt in Galveston, seen here in May 1945. (DND HS 1355-5)

was assigned to the officer training establishment HMCS *King*'s to serve as a sea training ship. The first *King*'s officers reported aboard on 14 September and two days later Lt C.C. Love, who had taken over briefly from Lt Hicks, handed over command of the ship to Lt Cdr Ben G. Sivertz, RCNR.

Sackville in May 1945, starting her career as a loop layer. Apart from the removal of her type 271 radar lantern, replaced by the tiny American radar antenna at the top of the foremast, her mid-ship section is largely unaltered. The main gun position forward has been replaced by a winch, and she has stepped a mainmast aft. (DND HS 1355-21)

THE END OF ONE WAR AND THE START OF ANOTHER

May 1945: the new foc'sle arrangment. Note the range finder just to the right of the funnel. (DND HS-1355-7)

As things turned out *Sackville*'s career as a training ship lasted little more than a month. In late 1944 the navy was looking for a ship to serve as a maintenance vessel for its network of indicator loops in the approaches to major harbours. This system of electrical cables recorded changes in the earth's magnetic field caused by the passage of a ship and were part of the port defences. The navy already had the old trawler *Rayon D'Or* under consideration for conversion to a loop layer, but an assessment of *Sackville* in October led to her selection for the task. Her remaining machinery was sound, and once her forward boiler was removed she would have a marvellous storage well for cables. It was a transformation that saved her.

Conversion to Loop Layer

Sackville began her conversion to a controlled loop layer on 9 November 1944 when she landed her guns and began de-storing. By the 15th she was reduced to care and maintenance status and in mid-December a floating crane came alongside to remove the foremast, main gun platform, oerlikon mountings and ammunition lockers. Meanwhile, dockyard workers removed the depth charge rails on the quarterdeck.

Number one boiler was dismantled and removed during the winter. The new cable tank could hold up to ten miles of cables of various sizes – and she would need all that space and more before her career as a loop layer was completed. By early March 1945 a dockyard crane was back alongside restoring *Sackville*'s upperworks. Sometime during this period a large steam winch was mounted where the four-inch gun once stood, two samson posts and derricks were erected forward of the bridge, a cable trunk was built down into what used to be the forward boiler room and two cable sheaves were fitted to the bow. In March the foremast was restored and a smaller mainmast added. The mainmast carried a large boom so a sail could be raised to steady the ship while handling cables. Oerlikon positions were rebuilt on the bridge wings and, following a brief docking, the two 20mm guns and the two-pounder pompom were re-mounted. Sporting a new all-grey paint scheme and "Z 62" as a hull number, her engine trials were conducted in Bedford Basin in April under her new captain, Lt J.A. Mckenna, RCNVR. *Sackville*'s conversion to a loop layer was officially completed on 21 May 1945.

It was, in fact, a good time for a corvette to become a loop layer. The war in Europe officially ended on 10 May. By June corvettes already lined the harbour at Sydney awaiting disposal, and more were arriving each day. By the end of July ninety-four of the surviving 113 RCN corvettes were decommissioned. Forty-five were sold for scrap, although in the end only forty-one

Sackville was almost certainly the only Canadian corvette ever to be fitted for a sail! In this case, it helped to keep the ship's head into the wind during lifting or laying operations. (DND HS 1355-15)

went immediately to the breakers. Most of those, however, were *Sackville*'s mates from the first building program: thirty-one of the first sixty-four corvettes built for the RCN were cut up right after the war. Others steamed on for many years after in different guises, some – especially the excellent late-war Castle Class corvettes – as warships. *Sackville*'s stablemate *Amherst* was sold to the Venezuelan navy, but she refused to go. She broke her tow cable and was wrecked on the Gaspé coast in December 1945 while en route to her new owners. *Moncton* was bought by a Dutch firm, rebuilt as a whaler and served until 1966 when she was scrapped in Spain. Very quickly every last corvette in the Canadian fleet – except *Sackville* – was discarded.

In contrast, *Sackville*'s new career was just beginning. On 10 June she steamed to Sydney to lift the cables in the approaches to that harbour. These were brought back to Halifax for storage on the 14th and then it was off to Mulgrave to remove the loops guarding the entrance to Canso Strait. These too, all five miles of them, were landed at Halifax. In early July *Sackville* set off for Saint John New Brunswick, for her first visit to the port since leaving for the war in January 1942. It took *Sackville* five weeks of intensive work to lift the six indicator loops lying off the port. While she was there at least one senior officer suggested that *Sackville* be selected for disposal as part of a breakwater. That drew a sharp response that she was much too valuable now in her new role. Certainly, she proved her usefulness and capabilities off Saint John in the summer of 1945. So much cable was raised that on two occasions *Sackville* steamed into the Bay of Fundy to dump the surplus. The work was not completed until 22 August 1945, when she set course for Halifax once again.

The Halifax indicator loop system did not require lifting, it required maintenance – the real reason for *Sackville*'s conver-

THE END OF ONE WAR AND THE START OF ANOTHER

Engineroom casing, after gun position and the mainmast, May 1945. Note the cradle for the carley floats and the way the mainmast is stepped. (DND HS 1355-16)

sion in the first place. After a quiet period in September the work started in October. Boiler trouble and a defective forward winch delayed operations in October. According to the log, in November the ship's bell was lost over the side when it became fouled in a line passed from another ship! Winter weather through December, January and February plagued further work. Storms undid much work, ripping away buoys attached to the end of cables and forcing recurring searches for the loops. Work on the windswept open decks was extremely difficult. *Sackville* had, with very few exceptions, served her few wartime winter months on the North Atlantic run which was further north but blessed by Gulf Stream waters. The coldest waters of the whole Atlantic war – save for the Russian convoys – lay off Canada.

For that reason the RCN had given priority in winterization to its Nova Scotia based ships. *Sackville* was not winterized and it hampered her cable operations in the winter of 1945-46.

By early March 1946 work on the Halifax indicator loops was complete and on the 6th *Sackville* lay alongside, inactive. On 8 April 1946 her wartime career finally came to an end when Lt Mckenna was appointed ashore and the ship was paid off. De-storing was completed on 17 May and *Sackville* was relegated to the reserve fleet.

The next four years were spent rather quietly in the reserve fleet. The ship underwent routine maintenance and inspection, including a short docking in the summer of 1947. Later that same year *Sackville* along with the frigates *Capilano* and *Wentworth* were pulled out of reserve, and their boilers fired up to provide steam for the dockyard during a coal miners' strike. When the miners went back to work *Sackville* and her sisters went back into reserve.

In the meantime, the RCN went through the post-war doldrums. The size of the fleet and the number of personnel in the navy reached their post-war low in 1947. In that year the RCN could do little more than keep a few destroyers running for training purposes. The fleet's aircraft carrier and two cruisers were largely idle, and most of the rest was in reserve. What saved the navy – and *Sackville*, too – from a short slide into oblivion was the Cold War.

Although the roots of the Cold War reached back into the Second World War, it was not until the Communist take-over of Czechoslovakia in February 1948 that tensions between east and west became untenable. The Western European Union, a military alliance, followed as did the Russian blockade of Berlin. The larger North Atlantic Treaty Organization was formed a year later, in April 1949, bringing together the WEU, Canada and the US in a common defence pact. Plans for an expansion of the Canadian

The final corvette form: the sleek and powerful Castle Class corvette *Coppercliff*. The Castles were an interim design between the basic corvette and the frigate, using a single corvette powerplant but adding fifty feet to a totally redesigned hull, topping it with a proper naval bridge and a fully modern suite of weapons and sensors. (DND CN 6133)

fleet were already underway when Communist North Korea invaded South Korea in June 1950. On 4 August, as the rump of the South Korean army and a few American troops clung grimly to the small perimeter around Pusan, *Sackville* was called out to act as the depot ship of the reserve fleet and to be ready "for loop laying if needed." In fact, the navy already had tentative plans to recall *Sackville* to service as an oceanographic research vessel.

Saved by the War – And a Change of Role

The need had been there for some time. During the war the RCN had discovered just how difficult it was to find submarines operating in Canadian coastal waters. It was generally assumed before the war that the sound beam of an asdic travelled in straight lines – like a flashlight beam. Wartime experience soon indicated otherwise. Changes in temperature affected the shape and direction of the sound beam. For example, a sharp change in temperature, a thermocline, could split the asdic beam and send it off in quite different patterns in the same water mass.

Canadians knew, for example, after the 1942 U-boat campaign in the Gulf of St. Lawrence that any submarine which got down into the colder layers of the Gulf was undetectable. The same held true for U-boats which managed to get down into the comparatively warmer deep water off the Grand Banks in winter: the warm water bent the sound beam back to the surface, limiting asdic penetration to about 200 feet. It was also very clear that bottom conditions played an important part in how asdic sound moved in shallow, inshore waters. Sandy bottoms absorbed or reflected the sound nicely, while boulder strewn bottoms sent it bouncing in all directions. The sound swirled around so much off Halifax in the winter of 1944-45 that some ships actually got asdic contact on their own hulls!

New equipment late in the war helped resolve some of these problems, but the real solution was a better understanding of the nature of the water mass and bottom over which anti-submarine ships were operating. Canadian scientists began oceanographic work related to anti-submarine warfare late in the war, including establishing a research station at St. Andrews New Brunswick, and the operation of a research vessel, *Ekholi*, on the west coast. However, there seems to have been little sense of urgency. When the old wooden schooner *Culver* was acquired from the British Royal Society in 1944 for east coast oceanography work and then found to be unsatisfactory, the attempt was simply abandoned.

Greater urgency seems to have governed the RCN's action by 1945. By then the old type of U-boats fitted with schnorkels were operating inshore with considerable success and impunity. The newer German type XXI and type XXIII U-boats were even better: fast, silent and extremely capable. There was no clear solution to these radical new U-boats. This was bad enough. However, the Russians had captured many of the new U-boats intact, as well as the building yards, equipment and designers. By the late 1940s it was estimated – wrongly as things turned out – that the Russians had perhaps 300 of the type XXI U-boats in service. It was time to figure out what was happening in the ocean.

The navy moved on the problem in 1945 by allocating the Bangor class minesweeper *Quinte* to the Naval Research Establishment (NRE). She was replaced a year later by the Algerine class 'sweeper *New Liskeard*, which carried the burden of NRE research for the next twenty-two years. Starting in 1947 the NRE, operating in conjunction with the National Research Council and the Fisheries Research Board, began an increasingly intensive program of oceanographic research. Until the late 1950s the work was co-ordinated by the Atlantic Oceanographic Group (AOG) at St. Andrews, NB. This work concentrated on basic physical oceanography and the geology of the continental shelf, all of which had both military and civilian application. AOG's initial work was done from the wooden minesweeper *Lloyd George* until 1950, when the trawler *Whitethroat* took over. By then the navy and civilian scientists were looking for a larger ship to conduct the research.

In the hustle and bustle of early Korean war expansion the navy was content to hold *Sackville* in reserve as a loop layer. Nonetheless, drawings were prepared by the Naval Research Establishment in early 1950 to construct laboratories on *Sackville* and convert her to oceanography duty. These were not carried out, but *Sackville* soon found herself engaged in oceanographic research. She underwent an extensive refit in 1951 to correct the defects accumulated after four years of idleness, and then remained – in readiness – in the reserve fleet until December of 1952. Early in 1953 she towed a number of old Bangor class minesweepers from Saint John, Halifax and Lunenburg to Sydney. Then, after a brief docking in Dartmouth, *Sackville* arrived off the Biological Station of the Atlantic Oceanographic Group at St. Andrews New Brunswick, on 6 August to load scientists and equipment for a survey of the Gulf of St. Lawrence. *Sackville*'s new life as an oceanographic research vessel had begun.

Off to war again, this time as an oceanographic ship. *Sackville* in 1954, little changed – except for paint scheme and hull number – from her 1945 appearance. (DND HS 29519)

CHAPTER 5
OCEANOGRAPHIC SERVICE

For a full thirty years, from that cold December day in 1952 when she was finally called out of reserve until a drizzly day in the same month in 1982, *Sackville* served the interests of science. Her usefulness and efficiency in that role is perhaps best measured by the simple fact that, of all the Second World War vintage ships which operated as research vessels, and of all those which might have been converted for the purpose, *Sackville* lasted longest. With some modest improvements she also managed to survive quite significant changes in the sophistication of modern oceanography. Although by the 1970s she was eclipsed by modern, purpose-built oceanographic vessels, *Sackville* retained her usefulness to the navy until she was over forty years old. Few ships have served Canada better.

Sackville's career as a research vessel divides nicely into three distinct phases. In the first ten years her work was balanced between civilian and military purposes, working for the Joint Committee on Oceanography. From the early 1960s to the mid-1970s she concentrated primarily on biological and geological work under the direction of the new Bedford Institute of Oceanography. In 1975 she went back almost exclusively to naval work, supporting modern research into underwater acoustics by the Defence Research Establishment, Atlantic.

Cold War Research

The process began in August 1953 when, as the Canadian Navy Auxiliary Vessel (CNAV) *Sackville*, with a civilian crew transferred from *Whitethroat*, she steamed to St. Andrews New Brunswick, to load scientists and equipment from the Atlantic Oceanographic Group. The AOG had been responsible for the navy's initial forays into oceanography late in the war, including basic bathythermographic (changes in water temperature with

Cruising northern waters in the late 1950s, *Sackville* now sports her familiar post-war hull number; a new mizzen mast designed to carry radio antennas, not a sail; a power workboat abreast the funnel; and her radar has been relocated to a pole mast atop the bridge. (DND PMR 87-58)

In 1964 the new laboratory, seen here across the end of the engineroom casing, was added: the first significant alteration of *Sackville*'s corvette profile. (DND DNS 29606)

increases in depth) and bottom surveys. During the 1950s AOG provided overall direction for oceanography in Canadian waters, continuing its own work in the nature of the water mass and the seabed. This was the RCN's contribution to the Joint Committee on Oceanography. Other agencies shared in the work and results. Naval Research Establishment (NRE) at Halifax conducted experiments in low frequency sound propagation and worked on a new bottom sound location system. The Fisheries Research Board participated in the general scientific research as well, and so too did geologists from the Department of Mines and Technical Surveys who took a lead in the basic research on ocean currents, bottom samples and seismic surveys.

All of this work was co-ordinated by the Oceanographer in Charge of the AOG. Thus *Sackville*'s movements in the 1950s show her a frequent visitor to St. Andrews, where she returned briefly to off-load in November 1953 after her first survey of the Gulf of St. Lawrence. In early 1954 she followed a regular shuttle between Halifax and St. Andrews conducting a winter survey of water conditions along the Scotia Shelf and the Bay of Fundy. As the weather warmed she moved to Sydney for a spring survey of the Gulf and estuary of the St. Lawrence River. The summer found her operating south of the USN's highly secret research facility at Shelburne Nova Scotia, doing "trials for NRE." These almost certainly were connected with calibrating the Shelburne array of the expanding underwater Sound Surveillance System. SOSUS was – and remains – a sophisticated system of underwater listening devices able to record the passage of vessels. The Shelburne station was passed to Canadian control in 1959. Once

Sackville completed her highly classified work in September 1954 it was back into the Gulf for more survey work in the fall, followed by another brief stint as a tug in December.

A refit at Lunenburg in early 1955 was needed, including a complete retubing of her boiler. Most of June appears to have been spent in St. Andrews getting ready for the year's work, all of which concentrated in the Gulf and Cabot Strait. *Sackville* sailed from St. Andrews in early July for Quebec City and then, following a brief return to the AOG base, she spent the month of August around the Magdalen Islands. Work in the Cabot Strait followed in September and by 1 October she was alongside at St. John's – the first time in over a decade. As before, *Sackville* ended the year towing, this time a scow to Quebec City, before returning to Halifax on 13 December.

The next year, 1956, was a hectic one as well. Oceanographic work off the Atlantic coast of Nova Scotia began on 11 January with another winter survey of the Scotia Shelf and Fundy areas. *Sackville* shuttled back and forth from Halifax to St. Andrews until late May. Then she steamed into the Gulf en route for Montreal for a brief layover. This much was simple science. But on 6 June she picked up an oceanographer from the station at Shelburne and made a passage to the largest USN base in the Atlantic, Norfolk, Virginia, to embark "special equipment" probably associated with the SOSUS system. En route her scientists took hourly bathythermographic readings. She was back in Shelburne by the 17th and then made two trips to Boston to embark more equipment. By the end of July *Sackville* arrived back in St. Andrews in preparation for another survey of the Gulf and the Scotian Shelf under AOG direction. Much of the next few months was spent in the Cabot Strait, with brief visits to the US base at Argentia (where there was a major SOSUS facility), St. John's and eventually to Gaspé, Quebec. *Sackville* was alongside at Halifax on 23 November 1956 after a busy year. She was now due for a lengthy refit.

There seems little doubt that *Sackville* was earning her keep by 1956. Scientists found her "very near to ideal for oceanographic work ... good seakeeping, satisfactory endurance and well equipped." The new "cathodic protection" (an electro-magnetic system for inhibiting rust developed by Canadian scientists at NRE) installed in the early 1950s also meant that *Sackville* had at least fifteen years of hull life left. The problem in early 1957 was that her one remaining boiler was shot.

Boiler Trouble and a Brush with Extinction

Refit and boiler problems kept *Sackville* inactive throughout all of 1957 and she might well have been discarded then. She was towed to Sydney in February and from there to Montreal for docking, and was back in Halifax by mid-June but her boiler troubles lingered. Scientists pressed for her to be converted to diesel engines so that these could be completely shut down when conducting very sensitive acoustic tests. *Sackville*, they argued, was needed "because of an anticipated requirement for increased effort in research into long-range detection techniques." This latter was almost certainly in response to the new and growing threat from submarines capable of firing missiles armed with nuclear warheads.

Sackville's lone fire-tube boiler needed replacement in 1957, but there was none to be found. The rest of the left-over wartime fleet was fitted with more modern water-tube boilers that operated with pressurized boiler rooms. There were no replacements for *Sackville*'s antiquated fire-tube boiler available, even within the reserve fleet. Re-engining with diesels was an expensive proposition, estimated to cost $1.068 million. This was much more than it had cost to build *Sackville* in the first place. In the

Above and right; *Sackville*'s final post-war form was achieved in 1968, as demonstrated here a few years later: fully modern bridge, cabins all along the waist of the ship and her foc'sle raised. (DND DREA 3437-1)

end, the need for *Sackville* was so great that her boiler was repaired. Ironically, when *Sackville* returned to DND research in the late 1970s her antiquated steam engines were found to be quieter – while operating – than modern diesel engines, something which made her very useful for towing acoustic equipment. For the moment, however, the advent of the nuclear propelled and nuclear armed submarine saved *Sackville* from the wrecker.

The next few years brought a slight change in the ship's operational pattern, as *Sackville* steamed into the arctic and made a number of trips to Bermuda. The arctic trip in 1958, made in conjunction with the research vessel *Vema* from Columbia University, was primarily intended (rather prophetically) to find new sources of fish should those on the Grand Banks fail. Voyages to Bermuda in 1958, 1959 and 1960 probably had something to do with mapping the boundary between the cold Labrador current off eastern Canada and the warm Gulf Stream waters that pass well offshore.

The time between these distant forays was easily filled by a busy research schedule in Canadian waters. *Sackville* was almost constantly on the move, conducting surveys, popping into St.

OCEANOGRAPHIC SERVICE

By the late 1970s *Sackville* was an essential partner to the government's new research vessel *Quest*, seen inboard of *Sackville* in this shot taken in the Halifax dockyard (note the submarine on the syncroloft). (DND DREA)

Andrews to land results and scientists, frequently visiting Halifax while in passing, and then back to sea. A short refit in Sydney in early 1960 was followed by surveys of the Laurentian Channel in June, Cape Sable and Bay of Fundy in July and Newfoundland and Nova Scotia waters in August. Then down to Bermuda again in late October, just in time for a hurricane, a brief stop in Halifax before going on to St. John's and then home to Halifax on 5 December.

After a long refit in Lunenburg much of 1961 was spent working for NRE on more acoustic trials southeast of Shelburne. That year *Sackville* also started her long association with the New Bedford Institute of Oceanography (BIO), opened in Dartmouth Nova Scotia, which began with trips into the Gulf and off the Labrador coast for biological studies.

During 1961 *Sackville* spent 101 days at sea, but 1962 was probably her busiest year ever as a research vessel. In that year she logged 21,000 nautical miles in seven major cruises that kept her at sea for a full 152 days. The first cruise took place in March south and east of Sable Island. This was the beginning of a Department of Mines and Technical Surveys (DM&TS) study of deep water circulation off the tail of the Grand Banks. *Sackville*'s scientists took water samples in sixty-one locations. In May bottom sediments were surveyed in the estuary of the St. Lawrence River. *Sackville* roamed from south of Sable Island to well up the St. Lawrence River in June and early July, obtaining data on current movements. This was followed by further DM&TS work on deep circulation south and east of Sable Island.

Then from mid-August to the end of September scientists from the Institute of Oceanography at Dalhousie University conducted the first major Canadian geophysical survey on the east coast of Canada. The same institute booked *Sackville* for a deep ocean cruise to Bermuda in October – at the height of the Cuban Missile Crisis. En route they studied "deep sea cosmic material" and plankton. The year ended with a twelve-day cruise to the Gulf to take water temperature readings as part of an ice forecasting project, and to sample marine organisms near the Magdalens.

Supporting Civilian Oceanography

Without a doubt 1962 was an exceptionally busy year for *Sackville*, and marked a shift in her work to largely marine biology and geophysical surveys for the next thirteen years. Perhaps because of the changing nature of her work (or perhaps because of the changing nature of the scientists embarked) operations in 1961-1962 revealed some shortcomings. Whereas a few years earlier scientists had found *Sackville* perfectly acceptable for oceanographic work, at the end of 1962 there were complaints about her suitability and standard of upkeep. The navy, it appears, had largely given up spending money on her, concentrating its funds on ships like *New Liskeard* and *Fort Frances* which were operating in support of NRE research exclusively on defence research projects. But *Sackville* remained a naval responsibility and even the Director of the RCN's scientific services complained about her condition in December 1962. No other Atlantic-based ship "employed for research has had so little money spent on it" he wrote. And yet there was "no comparison between *Sackville*'s sea time record and that of *New Liskeard* and *Fort Frances* put together." If *Sackville* was a more capable ship, he concluded, "our 1963 program would have called for an additional 10 weeks of sea time" – a nice compliment to the ship, although *Sackville*'s crew might not have liked the idea. It was necessary now "to equip her more fully for oceanographic research."

While senior officials battled over *Sackville*'s fate, 1963 proved a busy year, too. *Sackville* sailed on 16 January for the usual winter survey of the Scotia Shelf and did not tie up for the

Sackville and *Quest* alongside at Ponta Delgada, Azores, in August 1977: the old corvette's final visit to the eastern Atlantic (Courtesy Harold Merklinger, DREA)

season again until 29 November. During those months she plied the sea from the Gulf of Maine to St. John's and well up the St. Lawrence River: all familiar haunts. In the meantime the navy and scientists pondered what to do with her. Some thought she was now too small for the jobs required. The idea of replacement with a frigate was mooted, but given up when it was deemed too expensive to man and operate. In view of the heavy demand on auxiliary vessels for research time, and *Sackville*'s excellent hull condition, it was finally decided to find the money needed to bring her up to acceptable standards.

On 19 January 1964 *Sackville* arrived in Lauzon Quebec, for a refit and the needed additions. The most obvious change was the construction of a large laboratory across the end of the engineroom casing. An oceanographer's hut was also built at the foc'sle break on the starboard side. To improve her stability at sea a "flume stabilizing" tank was installed and, later in the year, so too was a vertical axis thruster – another propeller mounted in the bow. These latter two additions were attempts to ensure that she could hold a position, as steady as possible, especially when handling equipment lowered deep into the sea (in excess of 2000 feet). Corvettes were notoriously unstable to begin with, but building a laboratory on the stern and – later – a modern two-deck bridge simply made the problem worse. As things turned out, all efforts to stabilize *Sackville* proved worthless. Within months the ship's Chief Engineer complained that the stabilizing tank was not much use, and the bow thruster produced so much drag that it was removed in 1967. For that reason *Sackville* was limited to inshore work on the physical properties of water masses and on bottom sampling.

Sackville was ready for sea again by May 1964, and she spent most of the year, apart from the trip to Lauzon to fit the bow thruster in September, steaming off the east coast. The next year she undertook five major cruises. Four of these occurred largely in the Gulf: two on physical oceanography and two seismic surveys. *Sackville*'s principle task during seismic surveys was to act as the "firing" ship: lowering the charge into the sea and setting it off so that other ships and land-based stations could monitor the shock waves. Work was also done on physical oceanography for the Department of National Defence.

The improvements carried out in 1964, particularly the laboratory, were a great help but scientists continued to complain about *Sackville*'s limitations as a research vessel. Her station keeping remained poor, limiting the ship to shallow water experiments, and her range of a maximum of fourteen days at sea was short. Fortunately for her, though, the pressure on available ships was simply too great to let her go. All five ships attached to the BIO – *Hudson*, *Baffin*, *Maxwell*, *Acadia*, and *Kapuskasing* – were fully employed in 1965 and scientists were still short of time at sea. There was also a need for a ship to carry explosives, which *Sackville* was suited to do. Consequently, the Department of Mines and Technical Surveys, which operated the ships based at BIO, pushed in early 1966 to have *Sackville* modernized further. Although the ship itself was generally suitable, her standards of accommodation, messing and general habitability "remain that of the Second World War and are unacceptable by modern standards."

DM&TS now wanted a number of improvements made to see the ship through a further five or perhaps ten years. Accommodation needed to be modernized and expanded to house twelve scientists, washroom facilities brought up to date, a modern public address system installed, improvements in winches and equipment handling made, and safe storage for up to twenty tons of explosives provided. There were other needs as well, not least of which was a modern bridge.

Through 1967 the naval dockyard worked on preliminary drawings while the two government departments involved hag-

Sackville's final cruise came in July 1982 north of Bermuda, where the long ocean swell revealled that she had not lost a corvette's response to the sea! (Courtesy Harold Merklinger, DREA)

gled over who would pay. In the end it was decided that DND would continue to refit and operate the vessel for another five years, while the new Department of Energy, Mines and Resources paid for the proposed modifications.

Meanwhile *Sackville* soldiered on through 1966 and 1967, engaged in both civilian and some limited military research. Finally in 1968 she was taken in hand for her last major alterations. The naval style bridge, which had survived largely intact since the Galveston refit, was removed. In its place a modern two-deck fully enclosed bridge was built, altering *Sackville*'s look dramatically. In addition, the main deck between the break in the foc'sle and the laboratory aft was enclosed to add more accommodation. The ship that emerged from refit in 1968 was therefore a far cry from a corvette: only her hull lines revealed the truth.

A hectic schedule of research cruises commenced again in 1969, ranging from Newfoundland to Bermuda and throughout the Gulf of St. Lawrence. By all accounts *Sackville* was performing yeoman service, but that did not keep scientists from griping about minor if rather nagging limitations. In 1971, for example, they complained about the soot from *Sackville*'s funnel, which contaminated their samples. The navy, it turns out, was saving a few pennies by fuelling *Sackville* with a cheap grade of fuel oil. A change in fuel solved the worst of that problem. A few modern winches and pieces of special equipment were added and *Sackville* soldiered on. Despite the soot problem she was always a very clean and tidy ship, lovingly cared for by her civilian crew. And so, with the growth in university-based oceanographic research programs in the early 1970s and the resultant shortage of sea time for ambitious researchers, *Sackville* remained a good platform for most inshore tasks.

The navy agreed. As it concluded in 1971, *Sackville* had her problems but she would still be useful "for some time to come"

and was worth spending a bit of money on. The questions throughout the early 1970s remained; whose money and how long would anyone continue to spend it on a thirty-year-old ship? The navy retained responsibility for the annual refit and operating costs, while the Department of Energy, Mines and Resources, through the Bedford Institute, assumed the additional costs associated with research equipment. But the lines of responsibility were never very clear, and they were blurred by tightening budgets and increased requirements. The navy's own research program, under NRE's successor Defence Research Establishment Atlantic (DREA), was in need of more ships itself by the early 1970s. Its war-built fleet of research vessels was either gone or on it last legs: *Whitethroat* was sold in 1967, *New Liskeard* was scrapped in 1969 and *Fort Frances* was slated to go in 1974. *New Liskeard*'s replacement, the modern purpose-built oceanographic vessel *Quest*, was commissioned in August 1969. A similar ship designed to replace *Fort Frances* never got off the drawing board. By 1975 DREA would need a second ship to work alongside *Quest*.

Back to Defence Work

Through the early 1970s DREA made occasional use of *Sackville*, and in 1973 five cruises were made for DREA by the Pacific-based research vessel *Kapuskasing*. But *Kapuskasing*, too, was slated for disposal unless she underwent an extensive refit. In late 1975 it came down to making a choice between *Kapuskasing* and *Sackville*, although neither ship could fulfil the longterm requirements of DREA. As it turned out, the old corvette won the battle for survival. Although less manoeuverable at low speed than a twin-screw Algerine and a little noiser, *Sackville* was well suited for towing acoustic equipment and her deeper draft meant she was less likely to drift when trying to keep station. She remained notoriously uncomfortable in any sea, but those who sailed in her felt completely safe. And, as always, she was well maintained. As one senior DREA scientist recalled, "*Sackville*'s longevity relates primarily to a succession of proud and diligent operators." She was always "clean and bright". Over the next six years extensive use was made of *Sackville* by DREA as a second ship in the research cruises conducted by *Quest*: *Sackville* completed fifty-two cruises in all. During 1976 *Sackville* steamed 11,540 nautical miles, spending 90 days at sea. Much of her time was devoted to underwater acoustic work, often acting as the firing ship, discharging patterns of explosives while *Quest* monitored the rate and range of transmission of the sound through different water masses. Nine cruises were planned for early 1977, but five of them had to be abandoned because of the need for a lengthy refit. This resulted in a "serious disruption in the scientific program" for 1977.

The refit had its purpose, however. *Sackville* was clear of the navy's new syncrolift on 19 April, and within a month she was at sea with *Quest*. *Sackville* may have been old by now, but she was fit enough to steam with *Quest* for the Azores on 1 August. After calling at Ponta Delgada they pushed as far east as 14 degrees west – the farthest east *Sackville* had been since August 1944. The two ships covered over 5,000 nautical miles in one month, not bad for an elderly lady with one lung!

The pattern for the next few years was much the same. Her log often records *Sackville* "stopped and drifting" on a given spot of ocean, or largely stationary for days on end. Much of this work was associated with the launching and recovery of acoustic arrays and sound projectors, probably related to DREA's development of new passive sonar systems to counter increasingly more sophisticated submarines. Not all of these cruises were defence-related, however. Throughout her final years as a research vessel *Sackville* continued to support work on marine life, such as her 1979 and 1980 cruises to the Labrador Basin on sealing surveys.

Corvette watchers knew that *Sackville*'s days as an operational ship were numbered when the navy deferred her scheduled refit in January 1982. Without the needed work she could not steam for much longer than a year. *Sackville*'s last cruise came in the summer of 1982, when she and *Quest* conducted acoustic tests just north of Bermuda. Over a period of ten days *Sackville* steamed over 1,100 miles, but much of her time was spent with engines stopped and drifting while her acoustic projectors and arrays were deployed and their sound monitored. She returned to Halifax on 1 August 1982. When her engines rang off 1019 hours on that fresh summer morning her forty years of service to Canada came to an end.

Decommissioning

On 16 December *Sackville* made her final sailpast. Vice Admiral J.A. Fulton, Commander, Maritime Command, took the salute, joined by a small group of dignitaries. Nostalgic and curious onlookers lined the waterfront, standing patiently in the fog and icy drizzle as *Sackville* steamed past. Her long decommissioning pennant – the standard 380 feet for a warship – trailed a full ship length astern as she slipped by. And, just as on that long-ago rainy day in May 1941 when she first hit the water, ships in the harbour sounded their foghorns and whistles.

Her passing marked not only the end of her own service, but the final end of that remarkable fleet that fought and won the Atlantic war. For nearly forty years *Sackville* had steamed on as the sole survivor of 123 corvettes in the wartime fleet. Through most of those years she served alongside other Second World War veterans, many like herself serving in support roles. Most of these ships were disposed of in the 1960s. The final die-hards disappeared from Canadian service in the 1970s, with the last to go –

Paying off, 16 December 1982: *Sackville* is pushed past the dockyard, her long pennant drifting astern. (DND 82-5059)

A fine shot of *Sackville* on one of her last working days north of Bermuda in July 1982. (Courtesy Harold Merklinger, DREA)

the Algerine 'sweeper *Kapuskasing* – sunk as a target in 1978.

Kapuskasing's fate might well have been *Sackville*'s, but she had tricked fate before. The Germans had tried and failed to sink her several times. A fortuitous boiler rupture and conversion to an obscure auxiliary vessel kept her in the fleet after the war, the only Canadian corvette not discarded in 1945. Several times afterwards she came within a whisker of disposal. Perhaps the navy was looking out for her after all. By the time *Sackville* was retired from active service in December 1982 she was the last of the many – and for all anyone knew, the very last corvette. Fortunately for her, and for Canada, she would escape the wrecker's yard this time, too.

CHAPTER 6

RESTORATION

Throughout her post-war career *Sackville* was dogged by speculation about her imminent demise. One senior officer had recommended in 1945 that she be sunk as a breakwater. She was nearly discarded in 1957 when her boiler failed and the costs of fitting her with diesels looked prohibitive. Two major refits and additions kept her viable through the 1960s and early 1970s, although by 1970 senior officers talked about only four or five more years of service being likely. Yet she carried on far beyond what anyone could have imagined. In the end, *Sackville* was just too well maintained, too easy to operate and too useful to simply abandon. Her last lease on active life came when DND failed to find the money to build another research vessel like *Quest* to replace *Fort Frances* in the 1970s. *Sackville* had to stand in, and she did until 1982. By that time a number of groups interested in the fate of the old corvette were organized enough to save her from the scrapyard.

The Search for a Corvette for Preservation

Many people followed the fate of *Sackville* and the other wartime ships through the post-war period. Most of those retained and in some cases modernized for service in the Cold War fleet were discarded in the 1960s. The most successful anti-submarine ship in Canadian naval history, the River Class frigate *Swansea* with three U-boat kills to her credit, was simply sold for scrap in 1967. *Haida*, more representative of the professional navy's ideal but with her own remarkable history in two wars, was saved by a group of dedicated individuals. Sadly the nation took little interest and *Haida* was eventually taken over by the Ontario government. The little ships just drifted away into obscurity, discarded by a disinterested service, mourned by a few veterans and largely unremembered by the country as a whole. *Sackville*, a small ship from the reservist's war, seemed doomed to a similar fate.

There was, however, a growing realization in the mid-1970s that if a corvette could not be found soon and preserved, none would be left at all. While many minds across the country pondered the problem and the ultimate fate of *Sackville*, the Board of Governors of the Nova Scotia Museum began to act in 1975. Their interest stemmed in large measure from plans to build a new Maritime Museum of the Atlantic on the waterfront in Halifax. Such a museum would need a ship or two for display, and the Board felt that the new museum needed a strong naval component. The immediate problem was whether to obtain *Sackville* or try to find another ex-RCN corvette that was not so thoroughly modified. In 1975 they vacillated between the two options, passing a motion to work at preserving *Sackville* while beginning an active investigation to find one nearer to its wartime configuration.

By early 1976 it was established that there were at least three Canadian corvettes still in existence in South America. The Museum's Board of Governors soon fixed on two in the navy of the Dominican Republic, *Louisburg II* and *Lachute*. Both were late war "increased endurance" revised corvettes. *Louisburg* had the most interesting wartime experience. She became opera-

75

The initial hope for salvaging a corvette lay with *Cristobal Colon*, ex-*Lachute* on the left, and *Juan Alejandro Costa*, ex-*Louisburg II* on the right, of the Dominican Navy, seen here in 1976. They were destroyed three years later by hurricane David. (Ken Macpherson)

tional in time to participate in the D-Day landing operations in June 1944, followed by a stint escorting convoys in British waters before returning to Canada for refit in March 1945. *Lachute* barely made it into the war. She joined the fleet in the last days of 1944 and served as a mid-ocean escort until the war ended five months later. Both were sold to the Dominican Navy in 1947 and remained largely unaltered.

The prospect of acquiring a corvette in original configuration was exciting, and the Nova Scotia Museum soon focused on *Louisburg*. She seemed, on the basis of anecdotal information, to be in the best shape, she had the most interesting wartime experience and her name was appropriate for the province. In August 1978 the Advisory Council of the new Maritime Museum identified *Louisburg* as perhaps their "last chance" to obtain a corvette for the museum.

In the fall of 1978 arrangements were made for representatives from the Nova Scotia Museum, the Canadian War Museum and one from the Naval Officers Association of Canada (who also happened to work for a Montreal shipyard) to visit the Dominican corvettes. Everyone was anxious about their seaworthiness, particularly because it was known that they had not been properly maintained over the previous thirty years. The group visited in November and the news was not good. The hulls of both corvettes were seriously deteriorated and there was concern that neither ship would survive a long tow back to Canada.

Despite the troubling news plans were made to acquire *Louisburg*. Over the winter of 1978-79 the Nova Scotia Museum convinced the Canadian Cultural Property Export Review Board to support their scheme. The federal agency provided in the end $29,100 (US) to have *Louisburg* docked locally for a survey and $45,000 (US) to purchase it. The Nova Scotia Museum was to oversee the docking and any needed repairs, and send down its own people to do another inspection. It was agreed, however, that only when the full state of the ship was determined would a final decision be made. Many larger and potentially more expensive problems remained to be solved. So far no money had been found for towing, restoration or operation of the vessel.

With the commitment to dock and survey *Louisburg* and the knowledge that she was in poor shape, the Museum began to get a little anxious about the long-term costs of preservation and what it might do to a small museum budget. In May 1979 Mr. George Burns, former manager of Saint John Shipbuilding and Drydock, cautioned the museum on the perils of a thirty-four-year-old hull. Burns, as it turns out, had worked on *Sackville* and might have had a personal interest. But as the Board of the Museum noted, *Sackville* was in splendid shape thanks to years of devoted maintenance and NRE's remarkable cathodic protection system. Moreover, *Sackville* had a long and distinguished war record as well as three decades of post-war service to Canada. If the Advisory Council of the new Maritime Museum had any doubts about *Sackville*'s condition they were allayed in late June 1979 when they visited the ship. The Council was impressed by her "excellent state of preservation and remarkable appearance" both inside and out.

Meanwhile preliminary repairs were conducted on *Louisburg* in preparation for towing her into dock, which was supposed to be done in June. The docking was delayed until July and then put off again until 1 October. Then fate intervened. On 31 August 1979 hurricane David swept through the Caribbean, driving *Louisburg* ashore and smashing her weakened hull like an eggshell. *Lachute* was wrecked, too. Now there was only *Sackville*.

Saving Sackville

In the fall of 1979, then, the Nova Scotia Museum shifted its focus to saving *Sackville*, which was believed to have another four years of service left. This gave the Museum a bit of time to muster its case and find the needed funds. To ease the whole process an attempt was made to have Parks Canada declare *Sackville* a ship of national importance. This was refused because of the extensive alterations *Sackville* had undergone. Meanwhile, with the new Maritime Museum of the Atlantic (MMA) slated to open in December 1980 and no ship yet procured for the waterfront, the Nova Scotia Museum began to pursue other possible ships. The most promising was the CGS *Acadia* which dated from before the First World War. She, too, had a distinguished history, including armed service in both world wars. Nonetheless, the Nova Scotia Museum never lost interest in *Sackville*. When the aged corvette was docked in December 1980 and found to have at least another four years of life left, the new MMA repeated to DND its desire to add *Sackville* to the museum's collections.

In fact, *Sackville* had much less time left in service than anyone thought at the time, and the 'bidding' for her was starting to mount. While the Nova Scotia Museum worked away on the matter, the Naval Officers Association of Canada had its eye on *Sackville*.

The pace quickened tremendously in January 1982 when it was revealed that *Sackville* would soon be paid off. The Commander of Marcom intimated that *Sackville* would likely go to the Maritime Museum. The Museum itself remained concerned about the high costs of restoration and annual upkeep of the vessel, but was very anxious that these not be revealed. Other potential contenders for *Sackville* had now emerged, including the Marine Museum of the Great Lakes, the City of Collingwood, Ontario – where a large number of corvettes were built – as well as the Navy League and the RCN Association. It was clear by early 1982, then, that *Sackville* would be preserved by somebody, somewhere.

In February 1982 a special subcommittee of the Advisory Council of the new Maritime Museum was struck to tackle the *Sackville* problem. It included Vern Howland, who had led the NOAC group which was now working alongside the Museum to acquire the corvette for display in Halifax. The Museum itself, officially opened in April 1982, had secured *Acadia* and she was alongside. *Sackville* would add the needed naval flavour to the Museum's waterfront, but the costs of restoring, operating and maintaining the ship could not now be borne by the Museum alone. The subcommittee reported in May that restoration would cost about $1 million, and the annual running costs of the ship would be at least $60,000. These costs were based on rebuilding *Sackville* to her 1941 short foc'sle configuration, which was the initial objective of the committee. It was then agreed that the money could only be raised through a major fundraising drive and that had to be centered in Ontario. NOAC took a lead role in fundraising and in July the group began to put together a non-profit Trust to oversee the fundraising.

Meanwhile *Sackville*'s next scheduled refit was cancelled and she sailed on what everyone suspected was her final operational cruise in July 1982. As she lay alongside awaiting final disposal at the end of 1982 the Canadian Naval Corvette Trust [CNCT] was established with an impressive list of patrons and active members, most located in Ontario. When *Sackville* was finally paid off in December it was understood that ownership would pass to the CNCT. In the meantime the Trust consulted with Eastern Marine Services Ltd about the costs of restoration to her 1941 form. It was concluded that it would be easier and much cheaper to put *Sackville* back to her 1944 post-Galveston configuration. The Admiral in Halifax agreed and offered to help restore the external appearance of the ship quickly so she could be put on display during 1984 at the Maritime Museum.

In the event, things did not move that fast and the problem of ownership lingered through early 1983 as *Sackville* sat quietly alongside jetty 7 in the dockyard. The MMA still wanted the ship but the Nova Scotia government offered no money in support of the project. DND got cold feet over transferring ownership to a Trust which had no money. Most of this was resolved by April. It was agreed that the CNCT would assume ownership, the navy would provide ongoing support, including off-season docking and some maintenance, while the MMA provided the venue for displaying the ship during the summer months. It was not a completely happy compromise, but it proved workable. It was also agreed that the target date for full restoration to her post-Galveston refit state would now be May 1985, just in time for the 75th Anniversary celebrations of the navy.

Restoration

While the lawyers argued over the finer points of ownership, the CNCT Working Group in Halifax got on with the job of restoration. Eastern Marine Services Ltd was contracted to do most of the structural work, while Halifax Industries Ltd offered one week free docking to clean and close off the hull. In August 1983 the ship was transferred to the CNCT and the stripping of the laboratories, midships accommodation and "new" bridge began in earnest. These were pretty well bare by the time *Sackville* was docked, from 11 to 31 October 1983. Halifax Industries sandblasted her hull and sealed it under numerous layers of anti-fouling and anti-corrosion paint. *Sackville*'s cathodic protection – the electromagnetic system that had kept her sound for so many years – was modernized and increased to that used on a ship five times her size. All the underwater openings in the hull for cooling systems and vents were plated over. Her bronze propeller, which in conjunction with the steel hull and salt water turned the ship into a huge battery (hence the need for electromagnetic protection), was removed. By the time they were done it was expected that *Sackville* would not require docking again for seven years: all of this work HIL did at cost. When she emerged from the floating dock at the end of October *Sackville*'s machinery, cut off from its source of water, could no longer be operated. The heart that had driven her for forty years was finally silent.

It was while *Sackville* lay in drydock that the formal transfer of the vessel to the CNCT took place on 28 October 1983, with Senator Henry Hicks handing the ship over to the National Chair of the CNCT, the late Edmund Bovey.

By 31 October *Sackville* was back alongside jetty 9 and the process of dismantling the superstructure began. The laboratory was ready to go by 21 November and on the 27th a naval crane on a barge was moved alongside to lift the pieces free. The next day thirty tons of surplus anchor chain was also removed, leaving just enough to run from the anchors down into the chain locker for display purposes. Proceeds from the sale of the surplus chain and scrap metal were poured back into the restoration project.

Over the winter work moved apace on the forward portion of the ship and some restoration of wartime fittings began. In November and December the bridge and waist accommodation added to the ship in her 1968 refit were finally gutted. Meanwhile a wood and plastic sheet structure was erected over the after end of the engineroom casing and work on the two-pounder gun platform commenced. The latter was completed in February 1984 – the first inkling of what lay in store.

What really lay between *Sackville* the research vessel and *Sackville* the corvette was, of course, her massive new bridge and the accommodation along the main deck which so profoundly altered her profile. During February and March 1984 the bridge was slowly dismantled on the inside, including removal of an internal deck. When that was completed it was found that the 1944 bridge (less the wings for the 20mm guns) could be rebuilt

The process of restoration began in earnest on 11 October 1983 when *Sackville* entered the floating dock at HIL: note the long 'beard' of seaweed accumulated after nearly a year of inactivity. (CNCT Trust)

RESTORATION

Now without power, *Sackville* is nudged back to her dockyard berth by tugs 31 October 1983. Torch marks and missing fixtures are already evident on her laboratory. (CNCT)

Dismantling the laboratory, November 1983. (CNCT)

RESTORATION

The 1968 bridge ready for removal, May 1984: her new 1945 bridge lurks inside. Note the footplates for the 20mm mounting stacked in the foreground. (CNCT)

Above and right; The new 1945 vintage bridge revealed, 7 June 1984. (CNCT)

inside the outer shell of the 1968 bridge. This was done through March and April using steel donated by DOFASCO Canada.

By May it was time to remove the remnant of the modern bridge and the accommodation along the waist of the ship. The removal of the bridge called for some tricky cutting and lifting to avoid dropping heavy plates on the "new" bridge hiding within. It all went without a hitch on 8 May, and the crane returned on the 14th to remove the structures along the waist of the ship. With that done another thirty tons of weight was removed and the old ship's draft forward decreased by one and a half feet. The 'deconstruction' phase was over: it was time to start adding.

Work on the bridge wings and the housing for the type 271 radar began almost as soon as the last pieces of the old bridge were lifted away. Some of the leftover DOFASCO steel went into a new platform forward for the four-inch gun, while steel framing was added for the Carley float platforms. These represented the last significant steel additions to the ship. Much of the rest was a problem of 'fitting' out with the equipment and general paraphernalia found on a wartime corvette.

Some of this was easy to pull together, some had to be fabricated, some scrounged from far and wide, and some acquired with a bit of leverage and pressure. Although it was first believed that the Canadian War Museum had two four-inch breech-loading Mk IX guns, the Trust could find only three surviving examples in the world. Two were in British museums and the third – fortuitously – graced the lawn of the Amherstburg Ontario Legion. The Amherstburg Legion proved very reluctant to part with the gun but was eventually persuaded to do so in exchange for a World War II vintage four-inch CPR gun from the army museum at CFB Borden.

The Canadian War Museum (CWM) had one 20mm oerlikon gun of the proper mark, and the only hedgehog mounting still in existence in Canada. The CWM also had four inert bombs for the

Rebuilding the bridge, June 1984. (CNCT)

Cutting out the 1968-vintage passageway that covered the waist of the ship. (CNCT)

hedgehog. All of these were transferred to *Sackville* on permanent loan. The single 20mm gun and mounting were placed on the starboard bridge wing and additional dummy hedgehog bombs were fashioned from wood. A few years later additional 20mm guns were acquired and almost a complete set of authentic dummy hedge-hog bombs found. Acquiring a two-pounder pompom gun for the aft gun position was even harder than obtaining the four-inch gun. Only one example of the type was known to survive in Canada, privately owned in LeHave Nova Scotia. Protracted negotiations to acquire it for *Sackville* proved fruitless. It was not until 1986 that a Vickers Mk VIII pompom was tracked down and obtained as a gift from the Irish Navy: a very generous donation of an artifact that was destined for their own museum. The Irish Navy also eventually donated the type 144 asdic control equipment gear needed to complete *Sackville*'s asdic hut.

Proper depth charge gear was, by 1984, virtually nonexistent. Depth charges were largely abandoned after the war in favour of new ahead-throwing weapons like squid and limbo. Certainly the old Mk II throwers borne by *Sackville* and the other corvettes were discarded quickly. However, it turned out that the late war depth charge throwers, the Mk IV which retained the cradle, had been kept in storage and were resurrected in the 1950s as torpedo launchers. Some were still in service in the late 1970s and had only just been removed – it was just a matter of finding the new name for them in the inventory! The Naval Engineering Unit (Pacific) [NEU(P)] re-converted four of these throwers and had them shipped to Halifax. NEU(P) also fabricated fifty dummy Mk VII depth charges to fill the stern rails and line the deck beside the throwers.

Since *Sackville* was to be rebuilt as she appeared following her Galveston refit she also needed a number of Carley floats and a twenty-seven-foot whaler. None of these remained in the navy's stores either. NEU(P) undertook to build the whaler as part of its hull technicians training course. It was finished in November 1983 and was used to win the New Year's whaler race in Esquimalt before being shipped east. The west coast engineers also manufactured the ten- and twenty-man Carley floats needed to complete *Sackville*'s lifesaving equipment.

Ten- and twenty-inch signal projectors remained in naval stores, and these soon arrived. However radio equipment of the proper vintage was another matter. In the end the challenge was assumed by a group of amateur radio operators. They found the proper sets, installed them and fitted out the radio room. The radar systems were another source of concern, but here too volunteers came to *Sackville*'s assistance. Eventually the last known example of a type 271 console was located in England, from which sketches, photos and three dimensional measurements were obtained. The Combat Systems Engineers of the Maritime Warfare School and the Ship Repair Unit at Halifax built the type 271 and SW2C antennas, cabinets and instruments on a cost-recovery basis.

It was initially planned not to bother with interior restoration in phase I of the *Sackville* project, but the volunteers could not be restrained. The seamen's mess on the main deck was easily gutted by the local Working Group volunteers, while the Fleet School of Maritime Command offered volunteers to restore the area if the CNCT provided the furnishings. A drive was then started for hammocks, mess kits, seabags, utensils, cap boxes and the myriad of personal items that characterized the 1944 messdeck. Tables, benches and lockers all had to be fabricated.

Much of the work remained to be completed by the spring of 1985, but her hull and superstructure lines were fully restored and she sported a gleaming fresh coat of light disruptive camouflage. Although she looked perhaps a little too tidy and too clear of the ropes, cables, coke cases, timbers, storage lockers and

HMCS SACKVILLE 1941-1985

Starting to look like her old self, June 1984. The hull lines have been restored and the reconstruction of her wartime bridge is well underway. (CNCT)

The Amerstburg Legion's 4-inch gun – the last of its kind in Canada – in its new home, 15 November 1984. (DND HS 846302)

The 2 pounder pom-pom, courtesy of the Irish Navy, goes aboard. (CNCT)

Ready to go: *Sackville* off the Halifax waterfront, 3 May 1985. A few pieces remain to be found and fitted, like another 20mm gun and a 2 pounder for the after gun position, and she's cleaner than any corvette ever was, but the last corvette has been restored. (CNCT Lyncan photo)

paraphernalia that cluttered the decks of wartime escorts, there was now no mistaking *Sackville* for anything but a Flower Class Corvette again.

Recommissioning

On 4 May 1985 – the day before Battle of the Atlantic Sunday – the restored *Sackville*, "dressed overall for the occasion", lay alongside a special float in front of the Prince of Wales jetty in the dockyard for a dedication ceremony. Among the dignitaries braving the cold, windy spring day were the Minister of National Defence, the Vice Chief of the Defence Staff, the Commander of Marcom and members of the CNCT. During the brief ceremony *Sackville* was designated as the Canadian Naval Memorial and recommissioned as Her Majesty's Canadian Ship once again.

Bearing these honours HMCS *Sackville* was given pride of place in the International Fleet Review held in Bedford Basin in June to commemorate the 75th Anniversary of the founding of the Canadian Navy. A special trot buoy was laid for her and she was towed to her place of honour: the oldest warship in the review and the only veteran of the Second World War. The last of the corvettes and the last of the little ships that shaped the formative period of Canadian naval history had found a safe haven at last.

HMCS *Sackville* dedication, 4 May 1985. (Marcom Museum)

Appendix I: Canadian Corvettes

NAME	PENDANT NO.	BUILDER	COMMISSIONED	DISPOSAL DATE

Flower Class, 1939-40 RCN Programme

Name	Pendant	Builder	Commissioned	Disposal
Agassiz	K 129	Burrard	23.1.41	30.8.46
Alberni	K 103	Yarrow	4.2.41	sunk 21.4.44
Algoma	K 127	Port Arthur	11.7.41	6.7.45
Amherst	K 148	St. John	5.8.41	30.8.46
Arvida	K 113	Morton	22.5.41	30.8.46
Baddeck	K 147	Davie	18.5.41	30.8.46
Barrie	K 138	Collingwood	12.5.41	30.8.46
Battleford	K 165	Collingwood	31.7.41	30.8.46
Brandon	K 149	Davie	22.7.41	30.8.46
Buctouche	K 179	Davie	5.6.41	30.8.46
Camrose	K 154	Marine	30.6.41	30.8.46
Chambly	K 116	Vickers	18.12.40	30.8.46
Chicoutimi	K 156	Vickers	12.5.41	17.10.45
Chilliwack	K 131	Burrard	8.4.41	30.8.46
Cobalt	K 124	Port Arthur	25.11.40	30.8.46
Collingwood	K 180	Collingwood	9.11.40	30.8.46
Dauphin	K 157	Vickers	17.5.41	30.8.46
Dawson	K 104	Victoria	6.10.41	30.8.46
Drumheller	K 167	Collingwood	13.9.41	30.8.46
Dunvegan	K 177	Marine	9.9.41	30.8.46
Edmundston	K 106	Yarrow	21.10.41	30.8.46
Galt	K 163	Collingwood	15.5.41	30.8.46
Kamloops	K 176	Victoria	17.3.41	30.8.46
Kamsack	K 171	Port Arthur	4.10.41	30.8.46
Kenogami	K 125	Port Arthur	29.6.41	30.8.46
Lethbridge	K 160	Vickers	25.6.41	30.8.46
Levis	K 115	G.T. Davie	16.5.41	sunk 27.9.41
Louisburg	K 143	Morton	29.9.41	sunk 6.2.43
Lunenburg	K 151	G.T. Davie	4.12.41	30.8.46
Matapedia	K 112	Morton	9.5.41	30.8.46
Moncton	K 139	St. John	24.4.42	12.12.45
Moosejaw	K 164	Collingwood	19.6.41	8.7.45
Morden	K 170	Port Arthur	6.9.41	30.8.46
Nanaimo	K 101	Yarrow	26.4.41	30.10.45
Napanee	K 118	Kingston	12.5.41	30.8.46
Oakville	K 178	Port Arthur	18.11.41	30.8.46
Orillia	K 119	Collingwood	29.4.41	30.8.46
Pictou	K 146	Davie	29.4.41	30.8.46
Prescott	K 161	Kingston	26.6.41	30.8.46
Quesnel	K 133	Victoria	23.5.41	30.8.46
Rimouski	K 121	Davie	26.4.41	30.8.46
Rosthern	K 169	Port Arthur	17.6.41	17.10.45
Sackville	K 181	St. John	30.12.41	retained
Saskatoon	K 158	Vickers	9.6.41	30.8.46
Shawinigan	K 136	G.T. Davie	19.9.41	sunk 24.11.44
Shediac	K 110	Davie	8.7.41	5.11.45
Sherbrooke	K 152	Marine	5.6.41	30.8.46
Sorel	K 153	Marine	19.8.41	30.8.46
Sudbury	K 162	Kingston	15.10.41	1.11.45
Summerside	K 141	Morton	11.9.41	30.8.46
The Pas	K 168	Collingwood	21.10.41	24.7.45
Trail	K 174	Burrard	30.4.41	30.8.46
Wetaskiwin	K 175	Burrard	16.12.40	30.8.46
Weyburn	K 173	Port Arthur	26.11.41	sunk 22.2.43

Flower Class, 1939-40. Built in Canada for RN but retained by RCN

Name	Pendant	Builder	Commissioned	Disposal
Arrowhead	K 145	Marine	21.11.40	27.6.45
Bittersweet	K 182	Marine	23.1.41	22.6.45
Eyebright	K 150	Vickers	26.11.40	17.6.45
Fennel	K 194	Marine	15.1.41	12.6.45
Hepatica	K 159	Davie	12.11.40	27.6.45
Mayflower	K 191	Vickers	9.11.40	31.5.45
Snowberry	K 166	Davie	30.11.40	8.6.45
Spikenard	K 198	Davie	8.12.40	sunk 10.2.42

93

Trillium	K 172	Vickers	22.10.40	27.6.45
Windflower	K 155	Davie	26.10.40	sunk 10.12.41

Flower Class, 1940-41 Program
(short forecastle, water tube boilers)

Brantford	K 218	Midland	15.5.42	17.8.45
Dundas	K 229	Victoria	1.4.42	17.7.45
Midland	K 220	Midland	8.11.41	15.7.45
New Westminster	K 228	Victoria	31.1.42	30.8.46
Timmins	K 223	Yarrow	10.2.42	30.8.46
Vancouver	K 240	Yarrow	20.3.42	30.8.46

Revised Flower Class, 1940-41 Program
(increased sheer and flare)

Calgary	K 231	Marine	16.12.41	30.8.46
Charlottetown	K 244	Kingston	13.12.41	sunk 11.9.42
Fredericton	K 245	Marine	8.12.41	14.7.45
Halifax	K 237	Collingwood	26.11.41	12.7.45
Kitchener	K 225	G.T. Davie	28.6.42	30.8.46
La Malbaie	K 273	Marine	28.4.42	30.8.46
Port Arthur	K 233	Port Arthur	26.5.42	16.7.45
Regina	K 234	Marine	22.1.42	sunk 8.8.44
Ville de Quebec	K 242	Morton	24.5.42	6.7.45
Woodstock	K 238	Collingwood	1.5.42	27.1.45

Modified Flower Class, 1942-43 Program
(improved bridges, increased endurance)

Athol	K 15	Morton	14.10.43	30.8.46
Coburg	K 333	Midland	11.5.44	30.8.46
Fergus	K 686	Collingwood	18.11.44	30.8.46
Frontenac	K 335	Kingston	26.10.43	30.8.46
Guelph	K 687	Collingwood	9.5.44	30.8.46
Hawkesbury	K 415	Morton	14.6.44	30.8.46
Lindsay	K 338	Midland	15.11.43	30.8.46
Louisburg II	K 401	Morton	13.12.43	30.8.46
Norsyd	K 520	Morton	22.12.43	30.8.46
North Bay	K 339	Collingwood	25.10.43	30.8.46
Owen Sound	K 340	Collingwood	17.11.43	30.8.46
Rivière du Loup	K 537	Morton	21.11.43	30.8.46
St. Lambert	K 343	Morton	27.5.44	30.8.46
Trentonian	K 368	Kingston	1.12.43	sunk 22.2.45
Whitby	K 346	Midland	6.6.44	30.8.46

Modified Flower Class, 1943-1944 Program

Asbestos	K 358	Morton	16.6.44	30.8.46
Beauharnois	K 540	Morton	25.9.44	30.8.46
Belleville	K 332	Kingston	19.10.44	30.8.46
Lachute	K 440	Morton	26.10.44	30.8.46
Merritonia	K 688	Morton	10.11.44	30.8.46
Parry Sound	K 341	Midland	30.8.44	30.8.46
Peterborough	K 342	Kingston	1.6.44	30.8.46
Smith Falls	K 345	Kingston	28.11.44	30.8.46
Stellarton	K 457	Morton	29.9.44	30.8.46
Strathroy	K 455	Midland	20.11.44	30.8.46
Thorlock	K 394	Midland	13.11.44	30.8.46
West York	K 369	Midland	6.11.44	30.8.46

Modified Flower Class, built in Britain

Forest Hill	K 486	Ferguson	1.12.43	30.8.46
Giffard	K 402	Alex Hall	10.11.43	30.8.46
Longbranch	K 487	J. Brown	5.1.44	30.8.46
Mimico	K 485	J. Brown	8.2.44	30.8.46

Castle Class, built in Britain

Arnprior	K 494	Harl. & Wolf	8.6.44	5.9.46
Bowmanville	K 493	Pickersgill	28.09.44	5.9.46
Copper Cliff	K 495	Blythe	25.07.44	22.11.45
Hespeler	K 489	H. Robb	28.2.44	17.11.45
Humberstone	K 497	Inglis	6.9.44	22.11.45
Huntsville	K 499	Flem. & Ferg.	6.6.44	5.9.46
Kincardine	K 490	Smith's	19.6.44	5.9.46
Leaside	K 492	Smith's	21.8.44	19.11.45
Orangeville	K 491	H. Robb	24.4.44	5.9.46
Petrolia	K 498	Harl. & Wolf	29.6.44	15.9.46
St. Thomas	K 488	Smith's	4.5.44	23.11.45
Tillsonburg	K 496	Ferguson	29.6.44	5.9.46

About the Author

Marc Milner, a native of Sackville, NB, is a professor of history at the University of New Brunswick. He has written extensively on the Canadian Navy and the Battle of the Atlantic, including *North Atlantic Run*, *The U-Boat Hunters*, and *Corvettes of the Royal Canadian Navy* (with Ken Macpherson), and his novel *Incident at North Point*.

THE "SACKVILLE" TRUST

HMCS *Sackville* was built early in the Second World War, and has been lovingly restored and preserved as a museum ship. She is Canada's National Naval Memorial – a memorial to those who served in wartime, who so well established the traditions of loyalty and service to Canada; to those who serve every day in these challenging post-war days of peace; and to "all who pass upon the sea on their lawful occasions." The Canadian Naval Memorial Trust was established to ensure the memorial is available to all Canadians, and that the ship is preserved as a real symbol of Canada's naval people, past, present and future.

Like everything else that is worthwhile, preserving *Sackville* costs money, and will cost more as the ship gets older. While the Canadian Navy provides some valuable support, individual Canadians also contribute to her upkeep, and *Sackville* could not continue to survive without significant contributions from private and corporate donors. This support can take many forms – membership in the Trust is available for a small annual donation, or a more serious commitment can be made through the Trust Endowment Fund. Please get in touch with us for more information.

The Canadian Naval Memorial Trust
HMCS Sackville
P.O. Box 99000 Stn Forces
Halifax, NS
Canada   B3K 5X5